Blue Bel Air

By the Same Author

THREE NIGHTS IN THE HEART OF THE EARTH

BRETT LAIDLAW

W·W·Norton & Company
New York·London

Printed in the United States of America

First Edition

The text of this book is composed in Garamond #3,
with the display set in Discus.
Composition and Manufacturing by The Maple-Vail Book
Manufacturing Group.
Book design by Guenet Abraham.

Library of Congress Cataloging-in-Publication Data

Laidlaw, Brett.
Blue Bel Air / by Brett Laidlaw.
p. cm.
I. Title.
PS3562.A332B57 1993
813′.54—dc20 92-4631

ISBN 0-393-03406-2

W.W. Norton & Company, Inc.
500 Fifth Avenue, New York, N.Y. 10110

W.W. Norton & Company Ltd.
10 Coptic Street, London WC1A 1PU

1 2 3 4 5 6 7 8 9 0

For a home full of books

and curiosity, and the encouragement always

to be what we wanted to be,

this book is dedicated

to the memory of

my father

and

to my mother

Acknowledgments

I am grateful to the Bush Foundation of Saint Paul for support which greatly contributed to the writing of this book.
Thanks to Ed for nocturnal streetwise self-assurance;
To Steve for kindnesses great and many;
To John and L.A. for more fun in Menasha;
To Samantha for the house in Herbster;
To Lulu for the trout;
To Star for taking the long look;
And to Mary, for double happiness.

Part One

Alchemy

I went to college with a woman named Carla, who had majored in religious studies and was now a cook and worked at a small café on the University's West Bank. She had gone to high school with an odd and remarkable girl, a girl of prodigious, comprehensive intelligence, and though she had not known her well or been precisely her friend, the girl sometimes entered Carla's thoughts the way extreme and promising personalities from one's past will. She had heard that she had gone away to a prestigious school in the West, and wouldn't have expected that her destiny would lead her back to this city whose brightest lights usually gained their full brilliance elsewhere. One day a young woman came into the restaurant in the afternoon for a cup of coffee. Carla was on a break, sitting alone in the empty restaurant. She looked hard at the woman, and though she had certainly changed (she had been only seventeen when Carla had last seen her), there was something distinctive about her, a unique intensity in her focused attention, the cast of her green eyes and dark brows, her peremptory gestures in so small an act as pouring cream into her coffee. The woman poured cream, and Carla was back in chemistry class, botching her experiment in precipitates and eyeing with envious longing the green-eyed girl across the black counter who, though she was two years younger than Carla and the only sophomore in the class, had just produced, in what seemed to Carla sheer alchemy, the whitest crystalline floss in a clear liquid that brightened suddenly to bluest blue. (Carla's was a kind of muddy puce, and the addition of more solution turned the whole thing to sludge.) The woman in the restaurant opened a notebook; the girl in the chem lab turned her unfaltering attention to producing a chemical equation with the change of state precisely represented, energy conserved, every molecule in its place. Carla recalled how the girl would talk to herself in German as she wrote, taking a self-contained joy from the combination of expertise, from the paradox of those harsh mathematical sounds ringing out in her fine clear voice.

Carla was in the throes of chemical vertigo, recalling the terrible *mole* (it was the number of molecules that would fit in—a shoebox? A very big number), and stood a little dizzily to approach the woman's table. The name had come to her immediately: Sylvia Stenmark.

If she had recognized her in a crowd, or glimpsed her on a bus, Carla would probably never have spoken to Sylvia. But there was an intimacy in their sitting alone at separate tables in the empty restaurant that could not be dodged. Carla greeted Sylvia tentatively; Sylvia looked puzzled, then her face sparked with recognition. They talked for ten minutes—certainly the longest, maybe the only real conversation they had ever had. But their talk had the nostalgic intimacy of old friends regretfully parted, gratefully reunited.

Carla had to get back to work, prep the evening menu. There was a party next night, at the warehouse where Carla's friend Frank had his studio. Why didn't Sylvia come? They could catch up better there. Sylvia said she would, Carla didn't really believe it, but didn't discount it either.

Carla wasn't especially superstitious, she wasn't an astrology buff, didn't scrutinize tea leaves or invest in crystals (she did seem to feel that a *bouquet garni* of *herbes de Provence* held almost mystical properties); but she had a kind of faith that things happen for a reason, or can be turned to purpose. If the universe does proceed by invisible machinations, then coincidence, she believed, is its owner's manual. Anything uncanny, anything out of the ordinary, was worth considering and pursuing. This Sylvia Stenmark had commanded a unique fascination for Carla, a fascination she had never fully understood or reconciled; there were plenty of places to have coffee around the University; Carla saw this accidental reunion as anything but mere chance.

And I tend to believe that whenever you believe that something has happened for a reason, you'll contrive a reason and force it to size, even if, like Cinderella's stepsisters, you have to hack off toes and heels to make the slipper fit. But maybe our views didn't differ that greatly, except that Carla's was more optimistic. For some time I had been thinking, not in any rigorous way, about a motivating force I came to call *the secret desire.* It was something like the prime directive of one's existence. It was secret because it was too important to admit to oneself, and because, since it couldn't be followed by reason, nothing was served by knowing it. It was, I suppose, a variation on fate.

Carla called me to make sure I was coming to Frank's party. I had planned on it, and even though I assured her I would be there, she kept insisting I must come, because this girl Sylvia (she repeatedly corrected herself, throwing in "woman," to account for the five-year gap in their sketchy acquaintance) might be there, and I had to meet her. I perked up slightly at the mention of her name, because another Sylvia had recently popped up in my life, very tangentially, and Sylvia, though not rare, isn't that common a name. I took it as a minor coincidence and nothing more. I certainly wasn't going to mention it to Carla. She seemed to want my sacred vow that I would be at Frank's. I finally said,

—Carla, what is this, are you matchmaking?

A pause. Then she said,

—Well, I sort of hope she's already matched.

—Oh, I said, and felt stupid.

I should have known better, always should have. Maybe hidden desires make you hear what you want to hear.

I arrived at Frank's the next night without any feeling of momentousness whatsoever. It was the usual sort of studio party, with the usual crowd of artists, poseurs, hangers-on, the usual obscure music, the usual forced bacchanalian atmosphere. I had been a part of this group at the University, largely because of my friendship with Frank, my eternal roommate. I had graduated a year ago, having taken six years to get through, which made me, let's see, twenty-five (you'll remember how the mid-twenties tend to blur, you can't believe you're that old, then you can't believe you're that young). Now I looked on this crowd with a somewhat jaded eye; their pursuit of perpetual adolescence annoyed me, except when I chose to partake of it. Now I worked for a living—I was a reporter, photographer, copy editor, and whatnot for the community newspaper of the Red Earth neighborhood on the city's south side. I wasn't exactly climbing the corporate ladder, but since I routinely talked to the police, city council members, the mayor, "business leaders," and the like, I sporadically felt responsible, mature, and upstanding. But mostly I suspected that I had wandered into a masquerade without a mask.

In regard to my desires, my ambitions, my station in society, my life, I was, you might say, ambivalent.

Tonight I was highly ambivalent. I had intended to come to the party, but now I felt that Carla had forced me against my will. Frank was nowhere in sight, the beer tasted terrible, and as I looked around the room I realized I knew hardly anyone here, and didn't particularly like anyone I knew. The room was all concrete, with wired-glass windows up to the high ceiling on one wall, and it was loud. I didn't see Carla anywhere, and as I was coddling a bad attitude which was worsening under vigilant nurture, I thought I probably ought to leave. Carla would be angry with me, but I wouldn't endear myself if I waited around to meet her mysterious friend and behaved like a churl when I did.

So I was on my way out, but not soon enough. Carla's voice rose above the general din; I might have pretended not to hear, and made my escape, but I've never been very good at dissembling. I turned with my hand on the huge metal door, and let it clank shut behind me. Carla came beaming across the room with her friend in tow. The friend, this Sylvia, seemed to alternately pull back against Carla's grip and then spurt ahead with eagerness to immerse herself in the festivities. They came jerking along this way until they stood before me.

—Bryce, Bryce Fraser, Carla said, as if I might have forgotten my name. —This is Sylvia, Sylvia Stenmark. Sylvia, Bryce.

I looked at her and I blinked a few times as she held out her hand, and Carla began to look worried, till I took her hand, and I said,

—Sylvia, I'm sorry about the poem.

Carla gaped at me. Was this any way to greet someone, with a blanket denunciation of modern poetry? Was I drunker than I appeared? I was suddenly swept up by a delight that Carla couldn't understand, which was a shame, because if she had she'd have been more delighted even than I. By the time she did understand, it was too late for her to appreciate it. Sylvia was confused—her worst fears about coming to a party with a bunch of artsy weirdos were confirmed. And I, elated and transformed as I was, I would have relished the moment even more if I had known that it would be one of the few times I would ever have the advantage of Sylvia.

What Sylvia didn't know, for a few moments at least, was who on earth I was and what I was talking about. What Carla didn't know was that her Sylvia was my Sylvia—that is, the Sylvia who had made me prick up my ears when Carla mentioned her friend on the phone and this Sylvia who stood before me were one and the same. Big deal. Let me explain:

As junior staff member at the Red Earth *Gazette* (there were two of us, the editor Lillian and me), I had been assigned to edit the poetry quarterly, a two-page spread of sentiment and amateur art photography. One of the poems we received, titled "The Calculus of Imaginaries," had astonished me. It was about imaginary numbers, and it was a virtuoso effort, full of sparkling images and splendid metaphors, tightroping with considerable bravura a delicate line between sensuality and abstraction. It was the only good poem I had ever read in three rounds of editing the poetry quarterly. I loved it, it was so good it almost broke my heart. But I had to turn it down. We only run poems about crocuses and poverty, and the odd wry rhyming household ode. And our poems must be sincere, where this one hinted dangerously at irony. Printing anything truly poetic would have called our whole mission into question. The first time Lillian had compelled me to do this, she had promised that in the future we would switch off. She reneged, partly because she was the editor, after all, and really hated doing it, and partly because my father had been a poet, so I presumably knew something about these matters. But mostly, I suspect, she sentenced me to this punishment because she thought that reading fifty or so awful poems in search of the seven or eight most sincere might help me to develop some sincerity of my own. Ironically, sincerity is a trait which I possess in overabundance, but which I fight at every turn, for fear of the debilitating influence it can exert if not harshly checked. Selecting the prize poems was thus both easier and less morally restorative than Lillian might have hoped.

The poet of "The Calculus of Imaginaries," who took no prize, was one Sylvia Stenmark. The name had struck me as a pseudonym, I can't really say why. If I recalled the derivation of "Stenmark" correctly (I think I heard it in reference to a Swedish skier), it meant stony ground. This poet had chosen to till stony ground indeed, and had brought up roses. But not roses, exactly—maybe something more like pomegranates.

Ordinarily, the rejected poets received a simple thanks-but-no-thanks through the mail. Sylvia's poem had touched and excited me, enlivened a dull day, and I wanted to give her better than a rejection slip, so I telephoned. But writers, I forgot, never take rejection well, and are likely to become suspicious when you tell them their work is so good you can't print it. Once I got on the phone with her and realized my gaffe, I froze, and came out with dreadful editorial clichés like "very fine work . . . doesn't suit our present needs . . . ," the words going sour in my mouth. I had meant to speak of the exhilaration of finding this bit of gold amid the dross, of the calm beauty of her language, of the fine sensibility it evoked. While everything else was going bad on the phone, I also realized that had I said those things I would have been saying them for myself, to flatter my sensitive soul, to express its despair at being shackled in this impoverished world, forever yearning, forever thwarted in its quest for beauty and truth, and now, that's not becoming, is it? So I embarrassed myself, and I was both relieved and shattered when I hung up the phone. There was no reason to think that our acquaintance would continue.

And indeed, there was no reason for me to want it to. She had been brusque on the phone, as if I had interrupted her and idiotically so. Well, that much was true, but she might have risen above it. She had been rude. So what if I had turned down her poem? It wasn't because I didn't like it, but precisely because I *did.* It was just a stupid little community newspaper, and her poem would have been wasted there. Couldn't she tell we were kindred spirits, practically soulmates, both shackled in impoverished dross, etc.?

So when Carla had mentioned her long-lost friend Sylvia on the phone, the nominal coincidence had sparked a small irritation, the sting of remembered embarrassment. It had made me think vindictively of all the things I might have said to her, all the things I could say if I ever got another crack at her. . . .

And here she was. I was smiling what I thought to be a mysterious and knowing smile. The diverse points of our relationships had not yet triangulated for Sylvia, and Carla didn't have a hope. She looked both annoyed and afraid. Maybe this was the night when things got really ugly; maybe this was the night when Bryce's calm façade cracked wide open.

—What the fuck are you talking about? Carla said.

I waited for Sylvia to catch on. If I had to explain it would ruin everything that was building up so nicely. Sylvia's green eyes flashed.

—It wasn't that good a poem, she said.

—Oh, well, not that good. But good.

—I'm not that good a poet.

—Oh, I don't know.

—I'm good at a lot of things but that's not especially one of them.

—What all are you good at?

—I don't like to boast.

—Modesty is waived.

—Not that easily.

I had no idea where this was coming from, these lines dragged out from some romantic farce. The edge was real. It wasn't what I would have imagined as love at first sight, but there was a crackling electric line between us. Carla looked as if she might cry. Sylvia wasn't going to save her—she hardly knew Carla, who had dragged her to this bughouse and thrown her to this lunatic who fancied himself a poetry critic and verbal swordsman. I owed a great deal more loyalty to Carla. So I spoke.

—We know each other, sort of, I said.

—I can see *that.* Why didn't you tell me?

—I didn't know.

—You know her but you didn't know you knew her?

—Uh-huh.

Carla looked to Sylvia.

—Did *you* know *him* but didn't know you knew him?

—Well . . . What? Sylvia said.

So I jumped in here and explained about the poem.

—Bryce, you should have printed it, Carla said. Can you change your mind?

—Too late, it already went to press. Anyway, it wouldn't have liked its company. Like, "Sometimes from o'er my window sash, I see the flowers and the trash. . . ."

Sylvia had backed off a little. She was smiling. She looked shy. I couldn't figure it out. I liked her.

—I didn't even know you were a poet, Sylvia, Carla said. She was so amazing in chemistry class, she always knew exactly how many molecules were involved. She figured it out on her slide rule. She wrote it down. She's brilliant, she's going to win the Nobel Prize.

And Sylvia, to my amazement, blushed.

—I'm not brilliant, she said, shaking her head. —I'm not going to win anything.

—You won everything in high school. She did, Bryce, she won *every*thing. From tenth grade on.

I said:

—If you're no good at poetry, and you're no good at chemistry, what are all those things you're so good at?

She was at a disadvantage, she was truly embarrassed. But she looked me in the eye and said:

—Keeping my mouth shut, sometimes.

I dropped my head and said:

—Then maybe you can give me lessons.

Then none of us spoke, and the noise of the party seemed to rush in around us like nature abhorring a vacuum. Carla leaned toward me, and in a loud whisper, said,

—Be nice now, Bryce.

I said I would. Or, I think I said I would try. An awkward pause ensued. Sylvia seemed to have withdrawn, and so, I suppose, had I. The skirmish had been something extraordinary for both of us. Some mumbled remarks were exchanged, the blandness of which appalled us all, then Carla saw someone else she wanted to introduce Sylvia to, and she took her off across the room. As she was being towed away again, Sylvia gave me a sort of "Help me!" look, and I smiled.

Left alone again, I still could see no one I wanted to talk to. I still had seen no sign of Frank, the host. I watched Carla and Sylvia, and tried not to look as if I was watching them.

I watched Sylvia. She was pretty, I thought. Her hair was soft and dark, cut short. She was small, and her face was fine-featured. Her jaw culminated in a compact thrusting chin. Her nose was sharp and straight, and flared delicately. Her short-lashed eyes were a startling green. Already I had seen many of the faces of Sylvia—but then, I had provoked her. Why?

She didn't know what to do with her hands, which were short-nailed, and looked as if she used them. She put them in her pockets, reached up to sweep back her hair, clasped them behind her back, crossed her arms and tucked them away. Her attire was a sort of nerdy approximation of bohemian party-wear: navy blazer and white blouse with faded jeans and deck shoes. Her legs looked strong; she was small but not precisely petite.

She looked toward me and I started, felt goosebumps rise, dropped my eyes. I went out into the hallway. Frank's party had spawned adjunct gatherings in the studios along the corridor, and groups of people sat or stood leaning against the walls. I went down the hall to use the bathroom the squatter artists shared. My unease, my plain anxiety, told me I was fighting something. I wouldn't tell myself what it was. I stood a long time in the bathroom looking at myself in the mirror, or looking at the mirror but not really seeing myself.

As I came out of the bathroom I saw Frank at the other end of the corridor with a woman I didn't know. Probably he didn't, either. They stopped at the top of the stairs far down the bleak hall, kissed, and descended, their footsteps happily clattering.

I made my way back to the studio through a clutch of adolescent girl-poets dressed in black, shod in army boots, and sporting either great shanks of matted hair or no hair at all. They were being entertained by a man a few years older than I, whom I recognized as one of my father's former graduate students. He was telling them that the problem with our local writers was that they used too many short, gruntish Anglo-Saxon words, and not enough long, elegant Latinate ones. I thought that the problem, probably, was that they didn't know enough words, period, but I saved my pearls. And who was I to talk: I was the poetry editor who sought out the best poems, and then turned them down.

I had to push through this crowd to get back to the studio, and as I did I was almost overcome by the scent of the peculiar musk that was *de rigueur* among the matted-hair or no-hair crowd.

At the door I paused. I ought to go home, I thought. Frank was gone, I had put Carla off, I wouldn't know what to say to her friend. But I decided I should at least do Carla the courtesy of telling her I was leaving. So I hauled open the great door, an act which required all my strength.

Inside it was more crowded than before, but my eyes went straight to Sylvia as if she were decked with flashing lights. She was standing alone near the windows, by a table covered with bags of snacks and shards of pizza and empty beer glasses. She leaned against the wall attempting to simulate a posture of ease, then gave that up and stood shifting nervously, hands clutching elbows. She looked around, then looked down. She looked at her watch. Her expression alternated between sullen discomfort and a touchingly unconvincing effort at gaiety. I watched her for a minute from just inside the door, through a shifting screen of people. I was entirely charmed.

I went over to her. I said,

—Hi. Where's Carla?

She shrugged and said,

—I don't know. She left me with some people, and then they left too.

I asked her about Carla, about high school, about Carla in high school. Scintillating stuff. I didn't want to seem too personal or prying. She wasn't helping much. I didn't know what tone to take with her, nor she, it seemed, with me. Eventually we got on to what she did.

—I kill mice, she said.

And I thought, here we go again. But she was serious, always serious. She explained that she worked in a research laboratory in the bowels of the University hospitals. She was a lab technician on a project that had something to do with cancer, or diabetes, some major disease, and her main duty was as executioner of rodents. She took these mice, which had been encouraged to develop this grave malady, injected them with radioactive dye, dispatched the mice into eternity,* puréed them in a specialized blender, centrifuged the resulting liquid remains, and inserted the centrifuged remnant in a radioactive cell-counter. Report results to project head. Repeat. The work was part-time, but apparently quite involving. She was also still finishing her degree at the University. She had started college in California, and she had been at the University for three years, so she was twenty-two or so. She was thinking about transferring again, to the East, maybe. When she came back she never intended to stay.

*To kill a mouse, lay it flat on a table; press a pencil across the base of its skull (the mouse will appear to gaze up at you with loving but uncertain eyes); pull the mouse's tail.

—I could have finished this year, she said. But I kept dropping classes if I wasn't doing well.

—But Carla says you're brilliant. You won all the prizes.

She looked as if she would demur again, but as if that would be too much trouble she simply said:

—I am. I did win all the prizes. I drop classes if I think I'm going to get an A-minus, or a B-plus, god forbid. I'm a true compulsive obsessive. Some of the classes I've dropped, I probably would have gotten A's, but I couldn't risk it.

—Why?

—Why? Who knows. I might go to med school, is what I tell myself. And my mother would kill me if I brought home a bad report card.

—An A-minus isn't bad. B-plus isn't bad.

She gave me a slightly suffering look.

—I suppose you could guess that I know that, she said.

My damnable sincerity had done it to me again. For Sylvia, though serious, was quite ironic. She had gained confidence by catching me being patronizing, and now she wanted to know about me.

—What do you do?, besides breaking poets' hearts, I mean.

—I don't break poets' hearts; they break mine.

—Well it's mutual, then.

—You can't be serious.

—I only sent it because I was sure it wouldn't be rejected. It's just a crummy little newspaper. . . .

I took a small but genuine affront. Denigrating the Red Earth neighborhood *Gazette* was to me like using racial epithets: it's only all right if the speaker is included in the slur.

—Well I realize it's only a B-plus newspaper, I said. Still, it serves its purpose.

—No offense, she said.

Our conversation lurched along like this, from cordial to

sarcastic, occasionally verging on the intimate. I couldn't tell if she was really angry about the poem. But someone who drops classes unless she's assured of top grades must smart at any rejection, however small and meaningless. She never entirely relaxed—maybe because of the unfamiliar surroundings, and because I was on edge, too. But I also had the impression that she simply wasn't very accustomed to people. From time to time her embarrassment returned, for no apparent reason, as she glanced nervously around the room, and seemed to wonder, What am I doing here, with *people?* Can you trust these impressions? How could I have guessed that Sylvia was capable of altogether forgetting that a world without existed, could come up blinking from a solipsistic haze to thunder or a child's cry? When she spoke, in her high, ringing voice, any note of uncertainty vanished, as if each sentence were a weighty dramatic utterance. This combination of forceful competence and uncertainty appealed to me greatly. It made Sylvia seem like a brilliant troubled child, made me want to harbor her, shield her from the harsh real world. Fat chance. And it wasn't long before I saw the level of self-consciousness that accompanied this: Sylvia was well aware of her own vulnerabilities, and she would take care of them, of herself, herself; and so thank you very much, and go to hell.

Standing by the food table as we were, we both had been eating continuously, more to fill our nervous hands than to satisfy any hunger. And we had been drinking the bad beer. Sylvia abruptly set down her half-full cup on the table, as if she had just realized it was something disgusting. She pushed back a few damp hairs from her forehead. She looked pallid and flushed at once.

—I guess I'm about ready to go, she said. I'm not that used to drinking.

—Do you live nearby? I said. I can walk you home.

—East Bank. But I drove.

—I'll walk you to your car, then.

She didn't protest. I had thought she might. We found our coats in the pile in the corner, and went quietly down the stairs, Sylvia steadying herself on the railing. The stairwells had been decorated by a collaborative effort of the warehouse artists. The third floor showed an abstract splatter in primary colors; we wound down to diffuse, semi-human figures on a gray background at the second floor. A punky boy—male counterpart to the black-clad poetesses—was going at these walls with a black Magic Marker, adding definition to the diaphanous figures. He muttered something as we passed. "Fucking idiots"? "Fucking idealists"? Fucking something. He was pretty drunk. Next a dark red background on which cryptic motifs of fish, snakes, rodents, and insects floated, on the ground floor.

As we came out of the building, Sylvia said:

—When you called, I thought you were going to publish my poem. I've never had anything published before, never tried to. I didn't mean to be rude. I really was disappointed.

I nodded to accept the apology, almost began to explain myself again, but refrained. We walked the two blocks to her car as the sounds of the party waned behind us. At the car I said:

—Are you okay to drive?

—I'm okay, just a little queasy.

And that was goodnight.

Walking back to Frank's building, not sure that I wanted to go back to the party, I met Carla coming out.

—What happened to Sylvia? Carla said.

—She just left.

—Just left?

—Just now. You just missed her.

—I hardly got to talk to her!

—Sorry.

—You bring a girl to a party and she won't even dance with you.

—Sorry.

—Never mind. I've suffered humiliating rejection before. I'll survive.

—Sorry.

—Shut up. Let's have a drink.

—Good, I said.

I started up the steps and Carla said:

—But not there. Let's go to a bar. I'll go get my jacket.

I waited for her on the steps, and we walked to a bar. Carla bought me one. She told me about Sylvia in high school and how they'd met again at the restaurant. I bought Carla one. We stood each other another round each. Now it was getting late. Carla said:

—Are you in love?

—Am I in love?

—Are you in love? Are you smitten, or enthralled?

—Oh, I don't know. I only just met her.

—So?

—Smitten, maybe. I don't know. I like her, I think.

—He likes her. He thinks.

—But I thought she was your date.

—I thought so too, maybe. Never mind. I'm used to being bridesmaid.

She had drunk too much to conceal the truth she felt in that.

I walked Carla home. It had been a mild day, it seemed that spring might finally be arriving now in late April, but the night was cool, the air chill and damp. Carla took my bare hand in her mittened hand as we walked. We didn't have much to say now, but the silence was comfortable. Once she asked me,

—What are you thinking?

Nothing.

—Something.

—About your friend, I guess.

—We never were friends. You know more about her now than I do.

—Ah, but you *perceive* so much more. I can tell.

She squeezed my hand.

What Light Is Light

I had dark and thrilling dreams that night; for months I hadn't dreamed at all. Several times I rose toward waking through deep, gold-shot visions—it was exactly like rising toward the sun through dark water—rose toward the guttural screaming of cats under a swollen spring moon, in a brash and turbulent night; but I never breached the surface, descended again to dreams where the moon and the cats and the wind shaped impressions of bloody ripeness, the exhilaration and terror of life beyond reason. I woke near dawn with a sticky hot feeling all over me, but I remembered nothing of my dreams. I tried to sleep again, but the cats kept on (was it love or war? or did it matter?), and I lay in a daze where the waking and dreaming minds flowed together, and the cats' anguished yowling was the voice not of mere cats but of some creature that commanded all darkness. It didn't scare me; it was exciting in a way; and there was nothing I could do about it.

I woke again in the damp early morning to a cardinal's frenzied solo. My eyes as they opened caught the spark of red in the green mist, a male cardinal in a dying elm that arched over the alley. His frantic, unmelodious aria ricocheted off walls and windows—it was so loud, it was as if he were amplified. What was he yelling about? *This is* my *tree, this is* my *tree.* Or, *Come on girls, you know you want it.* And, well, they did. How could they help it? Then into my half-conscious, meandering mind popped a poem my father had written to help me remember a grade school presentation on birds. Cribbing from the encyclopedia, he had composed:

A bird is a warm-blooded
egg-laying vertebrate,
covered with feathers
and gifted with wings.

Body temperature two
to fourteen degrees higher
than that of the mammals
is that of these flyers.

That we could have printed. It was probably the only one of my father's poems that I knew by heart.

The bird squawked and warbled on, the final act in the night's burlesque. A bus roaring up the avenue drowned him out as it passed, and the sounds of traffic steadily grew, but the cardinal played on like the Titanic's orchestra. When I got out of the shower he was gone.

I sat at my window with a cup of coffee, looking out the building's back on a small parking lot and the alley. A large puddle in the alley told me it had rained in the night. Only then I remembered thunder in my dreams. The weather was so unstable, unpredictable in this season. Rival masses of air, the hot and the cold, the wet and the dry, clashed in the heavens. It could be ninety degrees one day, and then the skies would burst with strange, thrilling thunder-snow the next. The meteorologists might as well have packed it up till summer.

I smoked my first cigarette as I drank my coffee. I gazed at the puddle, whose dark surface reflected power lines, the cardinal's tree, and the gaudy marquee of a roast beef sandwich place across the street, shaped like a huge cowboy hat. A breeze came up and the puddle, obeying the imperative of water everywhere, made way, made waves. The images were still visible there, now undulant, now shattered. I don't know why this seemed remarkable. It was as if the particular were making a show of its universality, to tell me something. I blew on my coffee, watched the image of my squashed face scatter.

As the mystic torpor of my dark-dreamed night slowly faded, I began to feel a restlessness, some vague anticipation. It seemed to invade me through my nose, as the damp air held every slightest scent up for inspection. Through the cigarette smoke and coffee there intruded a sharp, ripe scent, like wet hay and piss—the smell of thawing earth, wet clotted grass, winter-rotted leaves. This wasn't the sunny springtime scent of lilacs or fresh-mowed grass. It was deeper than that, and far more potent; an exhalation from the earth's own funky, sweaty bed. It made me think of a warm, enclosed body's smell. I guess it was just the very stink of life.

In among these varied odors, the raw and the cooked, there lingered some scent from last night. Was it perfume, or something of her skin, her hair? Why did I carry it with me? Except to shake hands, we hadn't touched. If I tried to smell it, it was gone. And then returned, just the tiniest, maddening tinge, teasing the rim of my nostrils.

I went down to buy a newspaper. Just outside the door I encountered one of the feline combatants, or lovers. It was Jeoffry, the gray tortoise-shell cat that belonged to one of my neighbors in the building. He was lying in the parking lot, licking his chops and smacking his lips. His mouth was ringed with red, and I drew back, then started toward him, thinking he was spitting blood. Two steps closer, I saw that he was spitting out feathers—he'd been curtains for the cardinal. He stood, shaking away the last feathery bits, and then slowly, deliberately stretched, expressing full sensual joy in the redolent spring morning, minced a few steps toward me, stopped to stretch again, extended his neck, retched, and vomited on the blacktop. He flopped down to examine what he'd wrought. I made a noise of disgust, and walked quickly on to the corner.

Back in my apartment I flipped through the paper and idly began the crossword puzzle. 22 Across stopped me with a jolt: "Who is Silvia? What ____?"—*Two Gentlemen of Verona.* I dropped my pencil on the table and stared at the clue, at the half-filled diagram. I felt a tingly calm, and a warm, slow understanding rose in me. I asked the god of coincidence, who peered over my shoulder, how it was that he had this line into my secret heart, knew things about me even I didn't know. One Sylvia, two: okay, that could be chance. But three is the fairy tale clincher. Its portent can't be slipped. From *GNOSIS* I took an "I", an "R" from *RAIN,* an "S" from *SORE,* and from—a long one, "Bushman's relative," what the hell was that?—*HOTTENTOT,* an "H"; finally an "E" from *BED:*

IRSHE. "Who is Silvia? What *IRSHE?*" Something amiss there, I hoped, and as I checked back "precipitation" turned from *rain* to *snow,* giving me: "Who is Silvia? What *ISSHE?*"

Who is Silvia? What is she?

Good questions. But was I supposed to answer them, or was Fate merely being rhetorical?

I didn't know a thing about *Two Gentlemen of Verona,* except that it concerned one Silvia, difficult to define, and, presumably, two gentlemen. I had a few Shakespeare plays on my shelf, but they were mostly tragedies, no *Gentlemen* among them. The only possible help appeared to be a volume by William Hazlitt called *Characters of Shakespear's Plays,* in a lovely old Everyman's edition that I didn't even know I had—I must have swiped it from my parents, for its looks alone, during some past fit of bibliophilia. It took me several back-and-forth flips, scanning the page headings, till I found the chapter on *Two Gentlemen.* It covered a scant three pages. I read it with delight, and learned next to nothing about the play, and precisely nothing about Silvia. The article consisted mainly of a debate between Hazlitt and an earlier editor (Mr Pope, Hazlitt believed), on the merits of the play. Hazlitt: "[*Two Gentlemen*] is little more than the first outlines of a comedy loosely sketched in . . . yet there are passages of high poetical spirit, and of inimitable quaintness of humor." Pope: "This whole scene, like many others . . . is composed of the lowest and most trifling conceits, to be accounted for only by the gross taste of the age he lived in: *Populo ut placerent.*" Hazlitt: "The scene of Launce with his dog (not that in the second, but that in the fourth act) is a perfect treat in the way of farcical drollery and invention." Pope: "I wish I had the authority to leave them out, but I have done all I could, set a mark of reprobation upon them, throughout this edition." The only characters mentioned were the titular gentlemen, Valentine and Proteus,

Valentine's valet Speed, one Julia and her maid Lucetta, and, of course, Launce with his dog. Julia is in love, apparently with Proteus, but it seems that he spurns her, for she plots to follow him in disguise. Lucetta says uh-uhn, don't do it girl, to which Julia counters:

> The current that with gentle murmur glides,
> Thou know'st, being stopp'd, impatiently doth rage;
> But when his fair course is not hindered,
> . . . by many winding nooks he strays,
> With willing sport, to the wild ocean.[1]
> Then let me go, and hinder not my course;
> . . . Till the last step have brought me to my love;
> And there I'll rest, as after much turmoil,
> A blessed soul doth in Elysium.

1. The river wanders at its own sweet will. —Wordsworth.

Valentine also is in love, we know not with whom. He thinks his infatuation is a big secret, but to Speed it's plain as day: "You . . . wreathe your arms about like a malcontent . . . walk alone like one that had the pestilence . . . sigh like a schoolboy who had lost his A B C. . . . You are metamorphosed with a mistress, that, when I look on you, I can hardly think you my master." Oh, so it shows, does it?

There was a lot of talk about love, then, but no Silvia. Who is Silvia? What is she? A minor character, maybe, but one much wondered about. Certainly I wondered about her. I wondered about Launce with his dog in the fourth act. I wondered what it all meant. I knew it didn't mean anything. But the fact that I wanted it to mean something, was that a little elbow nudge from my secret desire? Or was it secret at all? A sort of Valentine's secret, perhaps.

Though it was Saturday, I had an assignment to complete

for the newspaper. Our regular man-on-the-street feature, "The Roving Reporter Asks. . . ." Another of the sour plums, like poetry editing, that Lillian bestowed on me with ulterior intent. She shunted this duty off on me because, she said, I had such a winning way with people. I think what she really saw in me was the seed of a reclusive crank; she pictured me years from now (but maybe not too many years) sitting unshaven in a rented room, eating pork and beans cold from the can, cursing at the television. So she would thrust me out among my countrymen, my neighbors, that I might grow giddy on society's warm sweet mead. Wrong again. Not that I was naturally gregarious or charming. But being forced to impose upon strangers for a quote and a photo in no way opens one's heart to humanity.

So, armed with notepad and camera (Lillian wouldn't like to hear that I thought of my respondents as the enemy), I went out unhappily to learn the answer to some pointless question that I hadn't even phrased.

But when I stepped outside, what a brilliant day I found. The sun was burning off the haze and drying up the puddles. Elms and lilacs seemed to have burst to full leaf overnight, and on southern exposures crocuses and daffodils thrust boldly up to the sun, and I was almost moved to poetry. This was spring then, at last, though there had been warm days before, and the only snow left was in dirty fugitive remnants of plowed-up piles. The last weeks had been days of dormancy, waiting to see which way the season would turn. Now spring was here in full extravagance, there was no turning back.

The question, then, the question. I still needed a question. It had to be general, and uncontroversial. If you assumed even the slightest knowledge of current events, you could ask until the cows came home before obtaining a handful of usable responses. I had learned to employ the magical query favored by TV journalists to coax a three-second response from anyone in any circumstance, the hero of the big game in the locker room, the family at the murder site, the homeowner in front of smoldering rubble: "How do you *feel?*" Almost everyone knew that. So the form was set, and the morning provided the content. The Roving Reporter asks: *Do you get spring fever? How does it feel?*

Excuse me, sir. . . .

—Hay fever? Terrible. Itchy nose, watery eyes. . .

No, no, sir—*spring* fever.

—Oh! *Spring* fever. Well, about the same, really.

Spring fever, to judge by the early responses, made people surly. There was a fair flow of traffic, but people brushed past me, wouldn't meet my eyes, or looked at me as if I were a dangerous lunatic, a panhandler, a politician. I needed just a half-dozen responses, but from an hour's asking I had only two, a kid and an old man, and as morning waned the promising crowds also dwindled. I idled on a street corner, discouraged. Maybe what I had wasn't spring fever at all. Maybe I should change the question. The Roving Reporter asks: *Who is Silvia? What IRSHE?* That at least might intrigue them. With my back against the drugstore's warm brick wall, my face bathed in sun, my mind wandered away from work, if you could call this work. Whom did Valentine love, that made him sigh as though he'd lost his A B C? Was it Silvia? Why would loving her make him pestilential, malcontent? Knowing Shakespeare, there was a rub. Julia's affection wandered at its own sweet will, unhindered, to the wild ocean. The wild ocean—it led me back to my dreams, those hot fluid visions.

The slap of my notebook on the sidewalk startled me. It had slipped from my hand. I bent to pick it up, looking around with embarrassment. I roved a bit, walking fast. The streets were almost empty now, but I couldn't give it up; we counted on the column inches, the goofy mugs and pithy wisdom of our neighbors. I was saved by a group of older women exiting a German restaurant. They were feeling spring fever in the best sense, as well as a number of daiquiris, by the smell of it. Most of them weren't technically residents of the neighborhood. They came in from the suburbs for the schnitzel. But one of them said she used to live near here, and they were all happy to answer, so, good enough.

When I finished with the ladies, promising I would send them extra copies of the paper for their grandchildren, I dashed back to the newspaper office to drop off the film and type up the copy. I had it in mind to make it to the library, which closed in the early afternoon. I supposed I ought to find out about Silvia. My journalistic and academic background made me a slave to research. I was never happier than when I was looking something up, even though I held really no faith in facts.

Lillian was in the office when I blustered in.

—Where's the fire? she said.

—What? No fire. Just, spring fever.

—Ah, indeed. May Day. Stomp the ground, rings the bells. Time to wake the earth.

Lillian, a handsome woman in her forties, was one of the few living humans still proud to be called a hippie. Her hair, brown streaked with gray, fell to her waist, and displayed remarkable and everchanging ingenuity in the art of braiding. She had always favored tie-dye, and was bemused to be fashionable again after all these years. She hated shoes, and suffered acute pedial oppression all through our long winters. I wasn't surprised to see her feet propped up on her desk, her tender, grimy soles breathing free, her toes twitching happily in the open air.

—There are health codes, you know, I said; a reflexive twit.

Just the usual reflexive chuckle in response. I sat down at the office's one computer, booted up, logged on, began to type. I began working fast and serious—there often was a real, old-time journalistic excitement in tapping a breaking story into the clacking plastic keys, even though we were a weekly, and would never really scoop anybody, even though we were just—what was it?—"a crummy little newspaper." But I hadn't typed far on the Roving Reporter's investigation into spring fever before the absurdity of all this impressive technology plied to the purpose of my inane questions and the responses in kind overcame me. I sat back and laughed.

—What's so funny? Lillian said.

Her very slight smile showed only mild inquisitiveness. We didn't share a sense of humor. She wanted me to be sincere and outgoing, after all. I shook my head, and said,

—Nothing. Everything. Don't things ever just strike you as ridiculous?

She had already turned back to her placid editing.

—Not that often, she said with a soft lilt. —Do they to you?

—Well, yeah. Not all the time. I mean, pretty often, I guess.

She looked benignly on my babbling. And I could see that, compared with her bedrock sincerity and calm, maybe I wasn't that sincere, after all.

I finished my typing and checked my watch. Plenty of time to get to the branch library before one. As I prepared to leave I asked Lillian:

—What do you know about *Two Gentlemen of Verona?*

—But soft, what light through yonder window breaks? That's all my Shakespeare. Why?

She was being modest.

—Nothing much. I'm interested in one of the characters.

—You ought to find yourself a real girl, Lillian said.

I left on that, and was several steps down the sidewalk before it hit me: How did she know it was a girl?

The library doors were locked. Saturday closing time was noon, not one. Coincidence, I said as I turned away, what does it mean? Will I only learn the story of Silvia when it's *too late?* My flippant questioning of Fate had this slight ominous edge: Was I building something up from nothing but a brief flirtation and a crossword puzzle clue that would grow rutted in its own course regardless of how reality pulled?

The way back took me past the neighborhood park, where the annual May Day celebration was in progress. Lillian had been literal: morris dancers all bedecked in ribbons and bells and funny hats were clogging the living daylights out of groggy Gaea. I wouldn't want to be wakened that way. But if that's what it takes. . . . On a stage across the park a reggae band blared happy, garbled sounds. A group of African drummers underscored it all with fast, insistent rhythm. There were people dancing everywhere. Cotton candy, Pronto Pups—"A Banquet on a Stick." Carnival games. I bought a hot dog and sat on the grass. Despite the sunny mild day, the earth still held a wintry chill. It took its waking slow, bells and clogs or no.

My building too was cold. The earth would have to take that airy warmth, then let it rise through concrete, brick, and plaster. I had a thought as I topped the stairs, and I knocked on the door of my cross-the-hall neighbor, the owner of the cat Jeoffry that had eaten the bird. I'd seen a lot of books in his apartment when we ran into each other coming and going. He came to the door looking disheveled and several days unshaven.

—Sorry to bother you, I said. Would you happen to have a copy of *Two Gentlemen of Verona?*

He stared at me with a kind of wild look in his eyes. Then he nodded.

—*Two Gentlemen.* I think I do, I think I do.

He went and rummaged in a corner where floor-to-ceiling bookshelves spilled their contents out in a heap on the floor. He tossed books aside with both hands. There were books everywhere. The cat lay on the windowsill. It turned its yellow eyes to me and blinked and yawned and closed its eyes.

—Your cat killed a bird today, I said. A cardinal.

He stopped rummaging and, still bending over, turned to look at the cat.

—Bad kitten, he said slowly, and it sounded more like praise than reproof. —Wicked, wicked kitty.

He tossed aside more books, then snatched out his hand and grabbed a thin paperback, as if it might try to escape. He brought it to the door.

—*Two Gentlemen,* he said. A minor play. Some sections of questionable authorship.

—What do you know about Silvia? I blurted.

—Silvia, the daughter of a duke. Lovely, but somewhat insubstantial. Loved by Valentine, pursued by Proteus, a rake.

—Thanks, I said, holding up the book as I turned away.

—You can keep that, he said.

—No, no, I just want to borrow it.

—No keep it, please. I'm trying to get rid of all these books.

He swept his arm back to display the book-infested apartment, and by the look in his eyes you'd have thought the books were some inexterminable vermin.

—Okay, I said. Okay, thanks.

I started away across the hall, and he watched me, nodding. As I opened my door he said:

—You don't have anything on Giordano Bruno, do you?

I offered a moment's expression of concerted thought, then said,

—No, I'm afraid I don't.

—Never mind, doesn't matter, he said, waving it off, and then he closed his door.

You can't really skim Shakespeare, and I didn't have the concentration to read it through. All I did was confirm what my neighbor had told me. Valentine loves Silvia, but Proteus throws a wrench in their romance, putting Valentine on the wrong side of Silvia's father, the duke, who banishes Valentine. I'd forgotten about Launce with his dog in the fourth act. I formed no opinion on the quality of the conceits. And really, if truth be told, I wasn't all that interested in the specifics of the drama. What interested me was how deeply enmeshed in this fuzzy-headed notion of coincidence I had become. I thought back to conversations I'd had with Carla, late nights in college when nothing in all of history was so important as what we thought about any damn thing, and the universe trembled as we shoved our half-fledged opinions out of the nest (and down below sat a cat called Experience, his appetite all whet). What Carla said about coincidence, I think, was this: It wasn't that our lives were ruled by any sort of mystical force, exactly, but that coincidence provided opportunity for action and connection, it highlighted passages in one's life, so to speak. And that was fine, I suppose anyone would agree that far. But then what? You've noticed this coincidence, now what do you *do?*

I dialed Carla's number, and she picked up.

—I wanted to ask you something, about coincidence, and about Sylvia, I said.

—What a coincidence, Carla said. Sylvia is right here. Do you want to ask her?

Panic hammered my brain, and a sick, plummeting pang dropped through my gut. So much for abstraction, so much for assumption. Carla, I had forgotten, was looking for a real girl. Maybe Sylvia was too. I stammered and blabbed, I don't know what I said. I became aware that Carla was laughing.

—Relax, Bryce. You're not interrupting. She was just about to leave. I'll let you talk to her.

Hardly a moment to prepare, and then:

—Hi, Sylvia said.

—Hi, Sylvia, this is Bryce Fraser, Carla's friend, from the party last night.

Was that just last night? Couldn't be.

—Right, Sylvia said. I know.

Then I let silence speak for itself.

—Did you want to ask me something? Sylvia said.

—Well, I. Not really. I—

—I'm just leaving here. Why don't I call you when I get home.

I gave her my number. I would have perhaps fifteen minutes to regain some semblance of composure.

The phone rang like a fire alarm. After I landed I let it ring again. Two-and-a-half. It's not as if I've been sitting by the phone. No indeed. I've been pacing furiously.

—Hi, I said suavely.

—Hi, it's Sylvia, she said.

—Right, I know.

I could have slapped myself. She had every right to hang up.

—Are you still mad about the poem? I said.

—What poem? My belly's full and I'm happy. Poetry is irrelevant to the sated. Carla cooked me a wonderful dinner.

I agreed that Carla was a splendid cook, silently asking if she was going to spoil my broth. Sylvia asked again if there was something I wanted to ask her, and I had to try to explain that although, yes, Sylvia had prompted my call to Carla, no, I had not thought to find her there; that I had called Carla to talk about, of all things, coincidence. All these things happening at the same time, Sylvia the hub of proliferating spokes of chance.

—Well, you know Heisenberg, Werner, the physicist—father of the much misunderstood Uncertainty Principle. He might say that although things seem to be happening simultaneously, we can't say for certain that that is the case. So you can't talk about coincidence at all.

Heisenberg. I had been planning to check with him next. Sylvia added:

—I think it's just that randomness sometimes comes in shapes that we would call patterns. They're no different from any other group of points, or events, that the chaos machine throws out, only to us, observing them, they seem more ordered. It's more cognitive than phenomenological.

—Right, that's my point exactly, I think.

—I mean the order is in our heads, not in the things.

—I thought that was what you meant.

Still, I had to insist that it was a remarkable string of events, or anyway a humdinger of a cognitive chain reaction. Whatever it was, it gave one pause. My reading Sylvia's poem, Sylvia's walking into Carla's restaurant, all of us showing up at the party, my calling Carla and finding Sylvia, Sylvia turning up in the crossword puzzle.

—The crossword puzzle?

—Yes, in a quotation from Shakespeare. Silvia with an "i," but just the same—

—Ah, but you're forcing the data. Similarity isn't identity. Maybe even identity isn't. Assumptions are always the beginning of trouble.

How identity wasn't identity was over my head. I wanted to know more about this chain of seeming events—that is, Sylvia's life, I suppose. I asked why she had left California and that shining, vaunted institution that most of the students in my high school, a private prep school, would have given their eyeteeth, all their teeth, most of their limbs, to attend. But she had stayed only half a year. She just said it didn't work out, it wasn't how she thought it would be.

—I got into Yale, and Brown, and Swarthmore too.

—Wow, I said; those names still held a lustrous, magical glow.

—I mean, I'm not trying to brag. I just did, I applied and I got in, that's all. But I wanted to go west. It seemed bright and fresh, and full of promise. That's an old mistake, isn't it, looking for the promised land. I should have remembered *The Grapes of Wrath.*

—*Day of the Locust.*

—Right. I was the first one in my family to go away.

—And?

—And you can never escape, can you?

—You said you wanted to try again. East this time?

—I guess I'll try.

I felt a sudden sadness and loneliness at the very thought; but I reckoned it was too early to plead with her to stay. But maybe if I kept her on the phone for all time. . . .

We talked for some time. Several times I heard other voices, footsteps and doors. There were long pauses, and lapses into awkward formality, as at the party. Several times I asked if she needed to go, if I was keeping her from something.

—No, it's just, I'm talking in this hallway, the house phone, and, I hardly ever use it. It's just sort of odd. But no, I want to talk.

When my voice was about worn out I said I wanted to see her. She said when, and I said, tomorrow.

—I have to go to my grandmother's tomorrow, but you can come with, if you want.

I gave her directions to pick me up. She had a car and I didn't.

When I hung up I picked up *Two Gentlemen* and opened it on the scene in which Valentine prepares to go into exile. No more will he see his Silvia, and maybe death is better than that:

> *What light is light, if Silvia be not seen?*
> *What joy is joy, if Silvia be not by?*

Oh. But that was Silvia, not Sylvia. And I was no Valentine.

Will of Weeds

At first sight Sylvia's grandmother's house made me think of those houses that show up on the TV news from time to time, where people have lived for years and years and never cleaned house, or taken out the trash, or been the least bit conscientious about the sanitation of their pets. When one day the squalor comes to light, then the children are taken away to foster homes, and the parents arrested and counseled, and the house, most often, must be burned to the ground. How does it happen? Things just start to pile up: the newspaper comes every day, day after day—what to do?; a bag of garbage makes it only halfway to the trash, and there becomes a kind of trash magnet, establishes hegemony in the hallway, and spreads. But doesn't it smell? You get used to it: putrescing chicken bones, moldy soup cans, dog shit—smells like home. It took so long for anyone to find out because the residents were always, oddly, scrupulous about their personal hygiene. "They seemed so clean!" Mrs Johnson from next door invariably exclaimed, while in the background a front-end loader carried away another festering heap, and that was how much she knew.

The house was a story-and-a-half white clapboard, begging for paint, and it sat on a slight rise surrounded widely by nothing, here on the rural fringe of the insatiable suburbs. The front yard was a tangle of tall grass and weeds. Small weedy trees and vines obscured the front of the house, made it look derelict, though gathered full draperies hung in the windows. A listing garage faced the side of the house, and an old blue car sat in the drive in front of it. Behind the car's rear wheels lay a scrawny dog. A crumbling shed stood a little distance off, and I sensed that hidden in the weeds lay old refrigerators, washing machines, and sundry other appliances, along with old tires breeding mosquitoes and disease, rusted fenders, transmissions, whatnot. And on the side of the house, spilling out through the breezeway door into the drive in a wide delta, as if the house had finally reached capacity and burst, were paper bags and boxes, and stray piles of household miscellany, and had this been some house you passed on a drive in the country, you would have sped by with a shudder, a shake of the head, thinking, *these people*. . . .

But this was the home of Sylvia's grandmother, an elderly German Catholic woman of high principles, I'd been assured, and though I hadn't quite puzzled out Sylvia's sense of humor, I didn't think she would joke about that. So we pulled into the drive. Sylvia turned off the car. We sat. I wanted to say something, my silence at this moment seemed a judgment, but I was dumb. In a moment we got out. As we did the dog sprang from beneath the blue car, and I jumped back behind the fender of Sylvia's car. Sylvia laughed.

—This dog is harmless, Sylvia said. This dog is demented.

The dog—a German shepherd, more or less, small and gaunt—crouched low with its ears laid back. It whined and squirmed, jangling its chain. Sylvia bent to pet it, saying,

—Kaiser, Kaiser, good girl.

But Kaiser wormed backwards beyond Sylvia's reach. Its hind legs wanted to approach, but its forelegs held it back. While its dirty paws tried the balance of fear and desire, Sylvia stood and turned away. She walked to the edge of the river of bags and boxes. Hands on her hips she looked into a box. She kicked it lightly. It clanked.

—Rummage, she said. Gran's organizing the church rummage sale.

I still stood uncertainly by the car. Sylvia motioned to me.

—It's safe, she said.

As I walked toward her I saw that the bags were filled with neatly folded clothes and linens, the boxes with several different sets of china, with arcane kitchen gadgets, tools, hardware, picture frames, vases.

—Rummage sale, I said. Why are they having it here?

—Oh, I'm sure this is just Gran's stuff. They'll have the sale at the church.

So the house had not burst from trash but was forcibly expelling all resalable items. I felt better. Among the loose items there were TV tables, a massive cabinet hi-fi, a dinette set, a nice pie safe. I looked at the small square house and thought it must be bare inside. We went up the breezeway steps made treacherous by cascading boxes. I heard the dog's chain sing, and looked just as Kaiser came to a violent halt at its end and fell flat and dejected in the dusty drive.

The breezeway too was packed with bags and boxes that rose

in towers, pressed against the windows, aspired to the rafters. Bing Crosby's blue eyes, timeless, a little attic-pale, peered down over the edge of a teetering carton of 78s. I stopped to admire an art deco floor lamp propped in the jumble.

—How much do you think . . . ? I said.

But Sylvia was on her way into the kitchen and didn't hear me. I followed her in. At the Formica kitchen table sat Sylvia's grandmother, looking like a rumpled sack herself. Her shoulders slumped in a baggy housedress, her feet in big black shoes were planted flat on the floor. One arm hung at her side, the other lay limp in her lap, and she looked as if she would never move again, would slowly go lower and lower like an eroding mountain until gravity pulled her into the earth, down to sea level. But when she saw Sylvia she began to regain a vertebrate shape; her shoulders slowly rose and straightened, like something inflating. Sylvia said,

—Hi, Gran, and bent to kiss her cheek, then introduced us: —Gran, this is Bryce Fraser, a friend of mine.

Slowly Gran held out her hand; more slowly still she nodded, as courtesy overcame fatigue and what seemed a deep reluctance to have her attention troubled with one more person, ever.

—This is a lot of stuff you've got here, I said.

—Everything goes, Gran said.

Her voice was quick, high-pitched.

—You can't sell the wringer, Sylvia said.

—Just collecting dust.

Sylvia put a kettle on the stove and measured instant coffee into mugs. As she did this she described how she had once nearly crushed her hand in the wringer's rollers. The machine had thus earned Sylvia's enduring devotion. Sylvia took an orange from the fridge, peeled it and sliced it into a bowl, sprinkled sugar on it, and brought it to Gran.

—Dust dust dust, Gran said, spooning up a bite of orange.

Sylvia and Gran caught up, and from their talk I gathered that Gran's church, Sacred Heart, was moving from its old—historical, even—brick home to new and modern quarters, a glass-and-steel spectacle with a freeway view and acres of parking. The rummage sale was billed as a fundraiser, but it was more a vehicle to involve the congregation, since the money they would raise that way would buy a couple of pews, or a stained-glass rendering of some plain saint. Gran, however, was donating her car to the effort.

—Sixty-two Bel Air, four new tires, new water pump.

She motioned out the window toward the blue sedan in the drive.

—Do you own a car, Brian? she said.

—Well, no, I said. But—

—Sheriff's deputy stopped here last week, Gran said. Says clean up the yard or he'll tag me. Says people are complaining. Who's complaining? Can't keep things up anymore.

The suburbs drew ever nearer; German and Czech farmsteads succumbed to duplexes and executive homesites. Industrial parks of corrugated metal fell from the sky at night, appeared as dawn visions to sleepy commuters. Sylvia brought three mugs to the table. She said:

—You haven't mowed the lawn in years. Why worry about it now?

Gran sighed. Her face was at once stern and softly fleshed. During pauses in the conversation she stared fiercely at nothing I could see, her lips set in a hard, down-curving line, as if she were considering something distasteful, reprehensible, even. As if every single instant of her life was serious.

—Ought to move, Gran said. Why can't people live and let live?

—That's it, Gran, strike out for the territories, Sylvia said.

—What harm is a few weeds? Health hazard. Your mother says sell the place, move in with her.

Sylvia laughed.

—Nice of her to offer though, Gran said.

Did the least hint of a smile soften her stern lips? Even I didn't like to think of her selling this place. I'd only been here ten minutes.

—There must be a lot of sentiment attached to this place, I said.

Sylvia turned sharply to me, shaking her head.

—My mother isn't sentimental, she said.

—*I'm* not sentimental, Gran seconded.

—Who here is sentimental? Sylvia asked, beginning to raise her hand.

I raised my hand as Sylvia pulled hers down, and I was voting alone. Sylvia and Gran both looked at me in silence. I lowered my hand to my coffee mug, lifted it, and drank.

Gran broke the silence, telling Sylvia that Father Moran had asked her to administer Communion at the weekday Mass. She was worried that she was not pure enough to handle the host.

—After all, it is the very body of Christ, no matter what these Johnny-come-latelys say.

Furthermore, she considered the father's offer a mere sop to keep her content through this upheaval. She had been the soul of that church for she couldn't even say how long.

—I didn't know regular people—what is it? laypeople—could give Communion, I said. Isn't it, what? a sacrament?, that priests have to do?

Gran turned to me:

—Fraser, is it? Scotch. Presbyterian.

I should learn to keep my mouth shut—Sylvia was going to coach me in that.

—Lapsed, I replied.

I recounted how my family's affiliation with the church had ended, when my brother was denied confirmation because he missed two classes for hockey practice. There had been a fight—my father enjoyed lost causes—and the church had taken us off the mailing list, a local alternative to excommunication, if Presbyterians even bothered with that. I tried to tell it in a comic tone. When I finished Gran nodded once, then looked away, out the window where the wild lot sloped down to the paved county road.

—At least you had some faith, she said.

She sighed.

—Apple trees are dead. Every last one. Squirrels.

We all looked. Beyond the drive stood several apple trees, twisted, gray, leafless in this shank of spring, their trunks obscured by thistle and weeds. Gran drained the last of her coffee, raised the orange bowl to suck the last of the juice.

—Going to die like a shrew. They just fall over dead, she said, and she slapped the table. —Running along, and drop dead. Can't keep things up anymore.

And as she said it she began to slump, to deflate; gravity was winning out. Sylvia looked at her with what I would call loving disbelief. And then, whether by act of human will or intervention of divine, Gran checked her fall, rose up straight in her chair—as straight as could be expected.

—Back to it, she said as she stood.

Sylvia got up too.

—What can we do to help, Gran?

I suppose I was a little charmed by her volunteering me so naturally, but I remained seated for the moment.

—Haven't looked in the cedar closet, Gran said. Lord knows what's in there. Junk. Dust.

—We'll take care of it, Sylvia said.

—Everything goes.

The door to the basement was off the kitchen, and as Gran opened it a draft of dank air spilled across the sunny room. Gran descended heavily the echoing stairs, and for some reason, or maybe none, I thought of that order of angels whose love of God is so great that they burst into flame as they near Him, and they sweep hot past His Tender Eyelids hissing, "*Holy!*" It was a kind of cycle, like rain: those angel ashes fell and at lower altitudes took form again, and were taken aloft on the updraft of divine love to blissful immolation once more. Unlike hydraulics, this system was perfectly contained, because God Is. Gran swayed down the dim stairs; I tried to imagine her flaming toward God. I saw armies of thick northern women flaming wave after wave toward God.

Sylvia sat down again. She looked across the table at me. She saw me with those green eyes. I searched the air for the scent that had dizzied me the morning before, the scent I identified with Sylvia. I don't know if I found it. We still were not at ease with each other, I knew next to nothing about her, I didn't know what I was doing here. But we were able to look at each other across a table this way, and not lower our eyes. And looking at her I had the sense, which I had also felt the night I met her, and maybe even had read in her poem, that she understood everything, but in some untranslatable way, and the only way you could begin to perceive what she did was by being with her, as if she were a splendid guide but absolutely worthless at giving directions. And this was why, I suppose—even though I had rejected her poem and broken her heart and treated her ill; even though I wasn't sure even if I liked her, or was smitten or enthralled or plain in love; even though I bridled against seeing the force of fate in meaningless coincidence; even though I raised my ambivalence against her with the full strength of my ironic sincerity—this was why, having met her only once and talked once more on the phone—this was why I was here at her grandmother's house on the dire edge of the creeping suburbs for no apparent reason. Why was that? Because for whatever reason or lack thereof, I wanted to know what she knew, I wanted to see what she saw. She was looking at me. She emptied her coffee cup and said,

—Shall we?

—Let's, I said.

We proceeded toward the cedar closet, upstairs. The whole house was infallibly clean—I was still half-expecting hordes of cats, ceiling-high stacks of newspapers, orange peels and crusts of bread. We passed through the living room, darkened by brown and yellow-flowered curtains. A needlepoint portrait in a gilt oval frame showed Jesus suffering the little children. It hung above a green sofa, the cushions sheathed in plastic, lace antimacassars draped over the back. On the coffee table in front of the sofa lay a scattering of knickknacks, more fodder for the rummage sale: a crystal bell, a pair of cock-eared cocker spaniels, a porcelain thimble (it had commemorated something, but the fine gold script had worn away). Sylvia picked up a statuette of a dancing lady, milky pink, full skirt of swirling china lace. She waltzed her through the air a turn or two, then set her down. She turned toward the staircase, sweeping out her arm in a motion of auspicious introduction.

—This is Hermes, she said. My childhood confidant. The only one who knows all my secrets.

This all-knowing Hermes was a statue, a yard or so high, who stood on the newel post. With his skin of darkly tarnished bronze he blended into the gloom of the living room. In his left hand he held a staff with two snakes wound round, and from the head of the staff sprang a flame of sandy glass. His limbs were youthful, nicely turned. He stood with left foot forward and weighted, right foot behind, heel raised, as if he were about to wing off to turn some intrigue or deliver godly gossip. I recalled that it was Hermes who accompanied dead souls to the underworld. Sylvia snapped on a light switch beside the front door and the flame atop Hermes' caduceus brightened. She climbed partway up the stairs to be on a level with her old friend, and began to criticize.

—Grandpa got him in an odd lot from an auction, she said. See here, his wing is broken—these angel wings, ridiculous.

Hermes' mode of propulsion was not the expected winged sandals or hat, but two wings that emerged from his smoothly muscled back; the tip of one was sheared off, leaving a rough edge.

—He's the patron of cheats and thieves, and of eloquence. He's also the protector of boundaries. That's why people put him on their newel posts. In ancient Greece it was just the head on a stone pillar, a kind of phallic thing.

Hermes had a phallic thing. He was naked, and anatomically correct. I was a little embarrassed, for Sylvia leaned on the banister with her head about that high. She was so accustomed to him, she probably didn't notice. Should I have been surprised that her professed soulmate was a well-hung little bronze immortal, three feet tall?

—Gran's always wanted to get rid of him. He's a pagan god, angel wings or no. But Grandpa wrecked the newel post when he put him in, so he stays. See here?

She traced her finger along the statue's base, where the wood was chipped and rough. Her other hand rested on Hermes' thigh, too close for comfort.

—Grandpa was a good carpenter, but also, he drank.

By now I had noticed something about Sylvia's voice, which was that it came in two registers, a high, ringing, dramatic voice, which I had heard that first night at the party, and a slightly lower, more relaxed voice, which I had heard for the first time as we talked in the car on the way to Gran's. The dramatic voice was also her telephone voice. The gradation was subtle, and I had only marked the difference when we arrived at Gran's and the dramatic voice took over. This was dramatic Sylvia all the way, as she talked about Hermes.

—When I was little and we came to visit, I spent a lot of time with Hermes. I told him everything, and I had him curse people for me. Because I was a very unhappy child. Persecuted. My cousins locked me in a closet, the one we're going to now.

Then she cocked her head and looked at me with a wide-eyed, confiding expression, and declared:

—Isn't that always the way with the other-worldly.

She looked at Hermes, and her expression changed again. Her face turned cross, the jut of her chin became exaggerated, her dark brows tightened. Some secret communication passed between them. Had Hermes been behaving badly? or not aging well? His face was really quite dull, idiotic, almost—a façade, perhaps. Sylvia flipped the light switch off. She turned to start up the stairs.

—Hermetic, hermeneutical, herm, she said as she started up. —Mercurial.

As if I needed an etymology lesson. Up she went. I saw that she had Gran's thighs—not thick on her—strong, shapely legs. She went up quickly, so lightly, even in her displeasure, and with those sturdy legs, she ascended. As I started up the stairs I gave a backward glance to Hermes. Winged but flightless, he was a symbol of something or other; packed with portent, he was mum.

Sylvia's buoyancy was contagious, and I went up quickly behind her. Rounding the turn where the stairs curved into the second-floor hallway, I pulled up short, for Sylvia had stopped on the top step and stood facing down the hall. I fell off balance and caught myself against the wall. Sylvia didn't notice. She stood looking down the hall, while I half-reclined on stairs and wall, and peered around her, trying to see what there was to see. Something in her posture and her silence told me to shut up—I was learning—and so I waited until finally she raised a foot, placed it on the worn floorboards as if testing iffy ice, and took three breathless steps: it held.

But would it hold two? I followed her in some real doubt.

There was a bedroom on either side of the hall, each long and narrow, outside walls shortened by the roof's slope. The hallway ended at a large, odd door, made of rough boards and painted white; it hung a little crooked on large metal hinges, and was closed with a wooden latch you lifted to open it. Sylvia now moved toward the door. It was only a few feet away, but she seemed to start and stop a half-dozen times before she reached it. She grabbed the handle and tried to lift the latch, and though the day was dry the latch stuck. She pulled at it again, but still it didn't give. She turned around and faced me. She held her hands palms up, then dropped them to her sides.

—It won't open, she said. Maybe Gran has something else we can do.

I looked past her pleading face to the door.

—Maybe I can get it, I said, and I stepped forward to move past her.

But she blocked my way and her face turned angry and she said,

—Are you going to open it with your dick?

And she turned and with the heel of her hand slammed up at the latch. And the latch flew open with a woody squeal. And Sylvia said,

—*There! Fuck!*

And the big door sprang open like a live thing, and a rush of hot, cedary air engulfed us, and Sylvia staggered back, and Sylvia gasped. She turned and shoved past me, clutching her hand. I ignored her and looked into the closet—hatboxes, suits, a lambswool coat, an old fur. Cardboard cartons and paper sacks. What did I expect to see?

—Jesus shit! Sylvia said.

And she struck the wall hard with the hand she'd already hurt.

—God damn it, she said, bending and holding her hands between her knees.

—Sylvia! I said.

Each burst of her profanity jarred me like an electric shock. In this place it sounded like blasphemy, like cursing in a church. I wanted to grab her and clamp my hands over her mouth. I waited for thunderbolts, for the floorboards to give way and drop us down to hell. Sylvia straightened up, pushed past me again, grabbed the edge of the door and flung it shut. I jumped away as the door slammed and then flew open again and shivered wildly on its hinges while Sylvia brushed by me again and sat down hard on the top step. I looked into the dark maw of the closet, from which the heated fragrant air continued to issue, as if a hot wind blew through there. Caught between the two of them, I almost wished the floorboards would drop out. I went to Sylvia, leaned down and said,

—Are you okay?

—Jesus. *Fuck,* she said, and stamped both feet on the stair, then closed her eyes and dropped her head down so it almost rested on her knees.

I sat beside her, and reached for the hurt hand that she still held between her knees. It was red and a bit swollen. I rubbed it gently. In a few moments she raised her head and opened her eyes, but she still stared hard at her knees, her brow tight, her green eyes fierce. Slowly her look softened and drifted, her hard hunched shoulders relaxed. And then she drew her hand away, but gently. She sighed.

—My aunts and uncles used to lock my mother in there, Sylvia said. I don't know why. No, yes I do. They did it for no reason. And later, I had heard these stories, so I was afraid of the closet, I was three or four, and my horrible cousins, they locked me in there too. I blame my mother, of course. If it hadn't happened to her it wouldn't have happened to me. And she's there to take it out on. Transference of aggression, like cats, when they're pissed off, they'll attack anything. I know it makes no sense. We're all just crazy.

She stood up and looked toward the closet. Her voice was matter-of-fact as she said,

—Although a lot of it is her fault.

She sat down again. We sat silently for a minute, Sylvia nursing her hand in her lap, looking straight ahead with an unfocused gaze. I motioned back toward the closet and said,

—Maybe we shouldn't—

And Sylvia jumped in, saying:

—No. It's okay. Why don't you start? I don't feel very steady yet.

—I don't know what's in there, I said.

—Everything goes.

The closet gaped behind us, and the sharp rich smell of the dry cedar surrounded us. Sylvia sniffed the air, and appeared paralyzed. I stood, and touching Sylvia's head as I turned, approached the closet.

No wonder no one was sentimental. I stood at the door a moment, peering down the dim shelves and racks. I could feel Sylvia's presence on the steps behind me. There were ghosts here, but they weren't mine, and so fearlessly I pulled the string to a low-watt light bulb, stepped inside, looked around.

There was nothing here but old clothes—hats, sweaters, suits, a fancy dress or two. But it was hot, and hard to breathe, and I wouldn't have wanted to be shut in there. I turned quickly; Sylvia still sat on the step. We all have these little fears, let's be honest. Somebody had a book of questions that are supposed to plumb your deepest feelings and thoughts. One was, What would you do if you were locked in a closet? I said, Yell, bang the door, sit down. So, I tend to give up. I tend to think, What's the point? I don't have anything to prove.

I started with a stack of hatboxes. There was a lady's black pillbox with a veil, and a child's bonnet, a brown fedora with a white soiled band. I felt strange, like a trespasser. This stuff was intimate and old, the stuff that earned mothproofing, a family album of cherished apparel. There was family enough under this roof, so why was I doing this? Everything goes: the theme of the day; yet I felt called upon to make judgments. Everything here meant something, and I didn't know what. How about this baseball cap, Milwaukee Braves, never-been-worn, packed in a shoe box, wrapped in shoe wrapping? It was beautiful, and I wanted it, but wasn't that odd? The tiara was junk, and didn't need cedar, so what was it doing here? How could I send these things away without knowing why? At the rummage sale nothing would mean anything. It would be quaint, vintage, a prize. But it would all be pointless. Maybe that was the point.

Grandpa owned a lot of suits for a farmer. I gathered an armful and carried them out. Sylvia didn't move. I took the suits into the guest room and threw them on the bed. Then, just for a gag, to lighten the mood, I took off my jeans and put on the top suit from the pile. It was a double-breasted gray pin-stripe, gangster style. It hadn't looked that big, or I hadn't felt that small, but the coat sleeves draped over my hands, and the pants wanted six more inches of leg. I shuffled into the hallway.

—Hey, I love these suits. Take a look. What do you think?

I lifted an arm from which a limp cuff drooped.

—In those days there were giants in the earth, eh? I said.

I couldn't keep from smiling. What's so funny about too-big clothes? Sylvia turned with an indulgent smile, and then the smile was gone.

—What a bastard he was, Sylvia said. Please take that off.

My smile vanished too. And I was about to do as told. Then stopped with something cutting on my tongue. Said nothing, did as told.

When I came out of the guest room, unhappy, Sylvia was gone. I sat on the top step now, and lit a cigarette. With no ashtray, I rubbed the ash into my jeans, feeling adolescent. The cedar closet was a presence behind me, as much as Sylvia had been, sitting on the step. The house was quiet. I leaned forward to look down the stairs, and at the bottom saw Hermes. How it must have irked him to hold that stilted pose, year after year. Hermes is the god of quickness, everchanging. But gravity pulled as hard at him as it did at Gran. This whole house should have collapsed into the cellar, entombed Gran there in an avalanche of rummage. What held it up? Angels, gods, and ghosts, perhaps. Now Gran contrived to dispatch her past to the glory of God and the good of the congregation. Good for her, but I didn't like being left all alone to sort through Grandpa Bastard's suits and big black shoes and dandy hats. I was glad to be back in my own clothes. I sat at the top of the stairs as if on a tiny island in a perilous sea. Still, mingled with my confusion and annoyance, there was something else—a faint excitement, intrigue. Did I still imagine I could save Sylvia, a damsel in distress? Why would I want to? Anyway, if the damsel wanted saving, she ought to sit still. She wouldn't.

My cigarette was down to the filter. No bathroom off the hallway, so I went through Gran's room and found one there, and flushed the butt. Gran's room was tidy and austere. Above the bed's plain headboard hung a crucifix, and Jesus again, solo. On the square dresser stood several photographs in gold and silver frames. Grandpa in uniform, tight at the collar, looking stern and German. His head was smaller than I would have expected. In his features I recognized Sylvia's chin, her sharp nose. In one family portrait I found Sylvia, a little girl, nine or ten years old, I guessed, with her mother and father and two sisters, one just a toddler, the other closer to Sylvia's age. I studied. I knew nothing about her family, except that she had sisters. And that there were problems. Which is always a good guess. Sylvia's mother showed the descent from Grandpa to Sylvia—fine features, a big smile and crinkly eyes as if she knew a secret. Sylvia's father was big and fair-haired, lushly bearded, in wire-framed glasses, wide-collared print shirt. He stood at the left rear, and Sylvia stood in front of him a little to the right. Sylvia held the baby on her hip, and her arm rested on her mother's shoulder. Her mother sat in a chair, erect and stiff, smiling straight into the camera. At her feet the middle sister sat with her legs curled under her and a hand on her mother's knee. There was something odd in this picture. Was it the lighting, or my imagination, or the camera that doesn't lie? The figures seemed odd and disconnected. They floated against the dark gray background, the spaces between them were palpable, as if some photographic trick had brought five figures from different negatives and combined them in one print. Sylvia stood in the place where I thought her mother should be.

I heard footsteps on the stairs, and a muffled thudding on

the stairwell walls. Sylvia appeared in the doorway, bearing boxes. I turned and faced her. She dropped the boxes.

—I apologize, she said.

She came into the bedroom, sniffing again.

—Gran hates cigarettes, she said.

—Sorry.

—Can I have one?

I gave her one. She saw that I'd been looking at the pictures, and she named her family for me. Mother Irene, father Ted, middle sister Rachel, baby Alison. Her parents were divorced, her mother remarried to a man named Richard. Her father was still single, as he would ever be, by the way she said it. What else was there to say? We went back to work in the closet.

The cedar smell seemed to thicken rather than disperse as Sylvia and I continued cleaning out the closet. There was a quality of warning to that scent now, as when you get a sniff of whiskey on the morning after a big drunk and know certainly and suddenly that it's poison. Still, Sylvia appeared calmer, in control. She took charge of sorting. She went through the boxes and scanned the racks like a sharp-eyed archivist. Of course not everything went. Gran's good wool dresses were here, and her daughters' confirmation gowns, and absurd and beautiful hats that the granddaughters would love. I thought of how I had blundered in here, with all best resolve, as if I could actually do this. Had Sylvia been testing me, to see how much I had figured out?

All of Grandpa's suits went to rummage. Sylvia merely waved at them, and let me pack them. But I dared to keep out one item from Grandpa's things, a boiled-wool overcoat, dark gray, with deep-blue lining. I asked Sylvia if I could take it. Maybe I was testing her.

—See what Gran wants for it. This is for the church, you know.

She spoke this dead poker-faced, the last part slightly mimicking Gran's tone. You see what I mean about sincerity. You see why I couldn't print her poem.

—The sense of smell is the most closely linked to memory of all the senses, Sylvia said. A few vagrant molecules, and zing, away you go.

We shoved the last boxes out of the closet. Sylvia closed the big door, set the latch. It was as if we were closing something in there, had just got out in time.

We took our boxes downstairs and crowded them onto the breezeway. I heard Gran pushing boxes around the basement. Sylvia took me out to show me the rest of the farm.

The dog came jangling, weaving, out from under the car. It stopped, ducked its head as if to turn around, then danced forward with a strange sideways gait, stopped and stood racked with indecision.

—There, there, Kaiser, Sylvia said. Weird dog. Good girl.

—Why is a bitch called Kaiser? I said. What's a female kaiser called?

Sylvia shrugged, bent over to right a tipped bag of linens. I looked at the car and wondered why Gran was selling it, since the new church would be farther away than the old. She drove to Mass every day in the blue Bel Air, Sylvia had said. Maybe it was like an old milk horse, and could only trace the familiar way to old Sacred Heart, refused to be steered from its patterned path. You got in and the car just went, but only to one place, soon defunct. Still.

—I want to buy the car, I said. Did she say it has new tires?

I was so pleased when Sylvia looked at me and had nothing

to say. At last she spoke, but she had waited too long and really shouldn't have said anything.

—The suspension's shot. It's not worth the price of insurance.

I would have walked over and bounced on the bumper, and proclaimed the virtues of springs and shocks, but Kaiser was in my path. I couldn't bear the thought of what my trying to get by would do to him. Her. Instead I turned and led the way toward the back yard.

Expecting more of the untamed spread in front of the house, I stopped in wonder at what I found behind it: a small, neatly tended yard running out square with the lines of the house, hacked out against the insistent will of weeds. The weeds held dominion over the greater expanse of the land that fell away down a long gradual slope. The way the weeds leaned in on the verge of this orderly patch created a tension all out of proportion to vegetable speed. Against the house on this southern exposure there was a flower bed with daffodils up and going, and tulips close behind; and in a tilled plot in the far corner rhubarb was unfurling fetal red, and chives and onion tops gave green relief. Across the back of the small lawn stretched clotheslines, sagging, the stanchions bending in as if in endless thwarted longing, where several blankets and bedspreads—more rummage—hung airing.

Sylvia walked past me and I followed as she crossed the lawn and pushed through the hanging blankets like the entry to a child's makeshift tepee. At the edge of the weeds Sylvia stopped and pointed. The land slowly dropped and then rose sharply to a road. Across the road the land had been cleared and graded, and skeletal houseframes rose above the black and sandy-gold earth.

—Grandpa used to farm all that, where those houses are, and more. He never wanted to be a farmer, inherited the land from his parents, but he kept buying more. Compulsive. What did he think, he'd turn into a lawyer if he farmed hard enough? That's what he wanted to be. Gran's been selling off the land bit by bit for years.

Sylvia moved into the rattling weeds, tore off a dry milk-weed pod and snapped it open like a fortune cookie to let the tufted seeds fly.

—Beware of people who don't like what they are, she said.

Sylvia had this way of speaking in pronouncements, with which I was uncomfortably familiar. My father used to do that. My father, also, was someone who didn't like what he was.

Sylvia took several long strides through the dry tangle, then sat. I settled down beside her.

—Grandpa beat the kids, beat the dog, probably beat Gran too, Sylvia said. She'll never say. He drank and dried out, drank and dried out. He was very pious when sober, and charming, intelligent. But he had a temper, you know—it runs in the family—and times were hard, blah blah blah. My mom's ten years younger than the next youngest, an accident. Grandpa was past his prime by then, one foot in the grave, so the other kids took charge of discipline. It must have felt good to give some back after all they'd gotten. The closet was just part of it. There were other things.

—What things?

—Other little tortures. They made her eat disgusting things, bugs, cow shit. I don't know. She was their little slave. It wasn't all bucolic splendor out on the farm. It was so enclosed, things could get pretty weird, I guess. What's normal is the way things are, right? Now they act like it was a joke, the things they did to her.

—Was there, was it—I mean, was there sexual abuse?

—I don't know. For the most part, that's a closed book. All the therapy we went through after the divorce, it's as if our deranged little family just dropped from the sky without precedent. It seems like—this sounds pretty trendy and pop—but if you can't look back to blame, then you look ahead. I don't intend to make that mistake.

—Was it all your mom? What about your father?

—My father! My father was the first in the family to go to college. Somehow, on a different farm, he became interested in—no obsessed, possessed by—theoretical mathematics. He was going to be a professor, a scholar.

—But?

—But me.

—What? You?

—Well, first my mom, and then shotgun wedding, and then me—I'm an accident too. And in the meantime, accounting.

—Oh.

—Oh. My mother, with her shining family example to guide her, what does she do? First chance, gets pregnant and marries a drunk. Classic. My dad, when he was sober, blamed himself. When he was drunk he blamed my mom. I know my dad is equally at fault, but in my heart of hearts I lay it all on my mother. Good little Electra. My mom, incidentally, for all her psychologizing, won't have anything to do with Freud. She thinks sex is beside the point. She's one of those women who uses it but won't admit it. In her own mind she's harmless, benign. More than that, she's always the victim.

Hidden down in the brown stalks, I felt safe and childlike, protected in this simple world from the evil power of grown-ups. Sylvia had collected a lot of burrs on her socks, and as we sat she pulled up her pant leg and began to pick them off. I reached over to help, but soon found my attention focused on Sylvia's leg instead—the pale stippled skin, sketched with a few fine hairs, the ridges her sock cuff had made on her ankle, over which I rippled my fingertips, her sharp straight shin, the soft pendulous curve of her calf which did not tighten as I shaped my hand around it. And then I thought of Carla, a winged thought flying in from nowhere; and I felt guilty and disloyal; but I did not take my hand from Sylvia's leg.

—Gran will never set foot in that new church, Sylvia said, playing with a cluster of round burrs that had discovered their mutual attraction.

—Why not?

—Just, she won't. She wants to die now.

—Like a shrew.

—Any loud noise will kill a shrew. Just clap your hands. They have an extremely high metabolism, high-strung, can't stand any shock. They must eat constantly, they scarcely sleep. Ounce for ounce, nothing beats a shrew for ferocity.

—So the old cliché is true, I said.

—What cliché?

—About a fiery woman like a shrew, high-strung, willful.

—I never heard that.

—Of course you have, *The Taming of the Shrew,* like that.

—Oh. I never thought about it. I think of rodents as rodents. You can't tame a shrew.

—I think that is somewhat the point, I said.

—Oh. Then I like that. That's good.

Maybe she would warm to *Two Gentlemen of Verona,* as well. Though she denied any identification with Silvia. She patiently picked one burr, then another, from her sock.

—Rats and mice have a memory of about one-point-five seconds, Sylvia said. Sometimes I think I'd rather be like that. No anxieties about the past. No family trauma. Fear and desire of an immediate and strictly physical nature. Easily dealt with.

—And constantly, and the same way, over and over again.

—Then everything is new, all the time.

—But you don't know it because you don't remember old.

—Well . . .

—The wonderful world of rodents, I said, more sarcastically than I had intended.

—Never mind, said Sylvia.

I sat up and shifted around in front of her, kneeling between her feet. I placed my hands just above her knees, and leaned into her. She didn't move until my chest was pressed to hers, and I knew I couldn't stop and I didn't, and she gave, falling back onto the trampled weeds, crushing leaves that smelled like faint dusty herbs. I held her behind the shoulders, pressed my face into her neck, kissed her throat. I felt her press one shoulder against me, and I rolled to the side but didn't let go. She put her arms around my back. Her breath, calm and even, warmed my neck. At first her hands lay slack against my back, then slowly her palms pressed, her embrace tightened. I hugged harder as she did, until we were pressed tight to each other in this warm hollow in the tall dry weeds, in the chaos that lay close about everything. There was nothing in this but the strange closeness of bodies, of distance dwindling, nothing more than I intended. Through the light fabric of Sylvia's shirt I felt the real, raw shape of her muscles and bones, sinews' live twitching, skin's shift; felt her pulse and breath under my hands, against my cheek, my neck. I thought I had never touched anyone so intimately before. Sylvia's small hands burned their shape into my back. I thought how those hands would coax a scared mouse from its cage, soothingly stroke its back, set it on a table, and break its neck. I closed my eyes, and everything was gone, there was just the two of us in this fierce embrace. We could have stayed that way forever, and no one would have found us.

We held it a minute, two. I opened my eyes. I knew she wouldn't let go. I had started this, and I would have to end it. But I couldn't. We did not relax, and we did not move. The world moves in this accidental way: action leads to action, forces play and counter and glance, things running into and off of one another. Now force was turned to force, and it felt more like a collision than a hug, and everything had stopped. A sharp stem pricked my side. I could imagine no end to this.

At last, without thinking, I rolled onto my back, pulling Sylvia on top of me, both still holding tight. My head was turned to the side, and her cheek was pressed to mine. We paused that way a moment, then rolled another quarter turn, onto our sides again, and as we did we both let go, and rolled apart—or rather, I threw Sylvia to the earth, and she allowed it, and she rocked away from me with the force of our parting, then settled and lay still. I couldn't believe how fast my heart was going, how hot and prickly my skin felt.

We sat up. Sylvia picked off a few more burrs, and pushed her pant cuff down. I breathed the dry weedy air, felt I was breathing crushed leaf and seed, and knew this smell always would signify this moment.

We stood and walked back toward the house, pushing two separate paths through the weeds. Sylvia walked around the side of the clotheslines, and I ducked and pushed through the hanging blankets, feeling almost as if I had to.

The house was still standing. I was a little surprised, had been imagining ruins. Gran was outside. She moved stiffly across the drive, sweeping the gravel with a broom and shouting at the dog.

—Lie down, dog! Gran said, jabbing the air with the broom. —Kaiser! Be still!

Kaiser wound herself up in her chain. The broom sent dirt and stones flying. Seeing her then, I realized suddenly that it was all a ruse, that she didn't want to die, like a shrew or otherwise, now or ever, and that furthermore, she never would. I had taken that angelic barefoot Hermes too lightly. All this grand relinquishing. But she had left the cedar closet till last, and then given the job to Sylvia. You can't get rid of everything. She still was in this world. You can summon your hardest will and say who needs an antique Boy Scout uniform, and laugh at kid gloves, silly prom frocks, tweed knickers. But then you reach in a box and bring out, say, baby shoes, and memory seeps from the scuffed soles through the skin of your palm, and you're done. Everything starts to shed its import like dye from cheap silk, and you're tattooed, tie-dyed, and drowned. Close the door, try another day, and for god's sake keep those baby shoes out of sight.

Sylvia went into the house. Gran dropped the broom against a pile of boxes. She stooped to pick up a twig, and tossed it to another area of the drive.

—Sylvia was showing me your land, I said. There's a lot. She said there used to be more.

—They'll take it all away, pay a tenth what it's worth, put up more of this ticky-tacky.

—It's a shame, I said.

—People have to live somewhere, Gran said.

I walked over to the car. The dog was out of sight beneath it, but its whimpering moan rose as I approached.

—How much do you want for the car? I said.

—Going to give it to the sisters, Gran said. Write it off.

—But I think I want to buy it.

—It's not for sale, worth more for the taxes.

—I'll give you five hundred. New water pump, you said?

—Rebuilt.

—And new tires? Five-fifty.

—Retreads. It's not for sale. Oh, you can have it.

—Five-fifty?

—No, just take it.

Sylvia came down the breezeway steps, the screen door rustling and clanking partway shut against a heap of bags and boxes. Gran turned to her.

—Your friend is taking the car, she said.

It sounded like: "Your *friend* just robbed a liquor store."

—You're really buying Gran's car? Sylvia said.

—No, she gave it to me.

Sylvia smiled.

—You could mow the lawn, Gran said with a backward sweep of her hand toward the brambled lot as she moved away toward the back door.

Sylvia walked over to me, hands in her pockets, smiling. She kicked a tire.

—Is it true about the suspension? I said.

Sylvia laughed.

We went back inside, and while I brought up boxes from the basement, Sylvia tried to convince Gran that she might need her car. Like everything else that happened here, the outcome was predetermined, the dialogue a ritual exchange. I only heard disjointed bits of talk between trips down the stairs.

—It's on the new highway, I don't like those new highways.

—You don't have to drive on it. There's a service road. The freeway's been there fifteen years.

I carried up a box of books. Sylvia motioned for me to bring it to her.

—Well, you'll need a car for shopping. For visiting the old people.

—Buying a new one. That old thing. Piece of junk.

By the time I had brought up the last box they had reached a decision.

—Keeping the wringer, Gran said. Don't know who'd want it.

Now who wasn't sentimental. Sylvia was looking through the box of books; the smell of mildewed pages filled the room. She kept one book and placed it in her lap as she rearranged the others in the box. She held it up for Gran and me to see: *The Children's Book of Favorite Saints.*

—Your mother's, Gran said.

Sylvia nodded. She flipped through the pages, turning the book to show me engravings of the sanctified, who gazed outward to the Great Yonder as they suffered their torments. Saint Sebastian was oblivious to the dozen or so arrows that pierced his limbs and torso, a blessed pincushion; Francis, with festering stigmata, preached to the birds behind a haze of red and blue crayon; a childlike Joan raised her sword and eyes to heaven as she burned. An innocent, wondering smile was fixed on Sylvia's lips as she looked at the pictures and read the legends, in enormous print, of her favorite saints.

We stood to leave, and Gran took Sylvia into the living room to ask if she wanted any of the knickknacks. Sylvia dinged the crystal bell, tried the thimble on her finger, and declined. She looked over at Hermes, who hovered in the gloaming. He refused to meet her glance, pretended he hadn't been watching. O ye inscrutable gods. I wanted to give him a good polishing, see how he felt then, all shiny and conspicuous. Hermes works by crooked ways, a soft word in the ear, a shrug, a calculated glance. Hermes carries tales, and pretends he doesn't care.

I remembered the overcoat as we were leaving the living room, and ran upstairs to fetch it. The hallway smelled faintly of cedar and cigarette smoke. From Gran's doorway I looked in at the picture on the dresser. If physicists could scan the cosmic mass of stuff and force in the instant before the big bang, they might tell the fortune of all time, the fate of matter and the void. What could I read in this picture of Sylvia's family in the calm before it flew apart? Could I trace the shape of what Sylvia was and would be in that odd photo? It was all there, surely. I tried to memorize it. I couldn't tarry long. I grabbed the overcoat from the guest bed and hurried down the stairs.

Sylvia and Gran were in the drive. Sylvia had coaxed Kaiser from beneath the car and was kneeling down petting her. The dog suffered her attentions uneasily. Sylvia was not much involved. Her face was stern, not joyful, the way it should be when you pet a dog.

—I'd like this coat if you're getting rid of it, I said to Gran. What do you want for it?

Gran took the folded coat and let its thick weight fall from the shoulders.

—Let's see it on, she said.

—Don't put it on, Sylvia said without looking up.

—Why shouldn't he put it on? Gran barked.

—Should I put it on? I said to neither in particular.

—Oh go ahead, put it on, said Sylvia.

I slipped into the coat as Gran held it. I loved the heft of it, the downward weight, the silken slide of the lining, the hem lapping at my calves. I felt safe in that coat. It was enormous.

—Too big, said Gran, reaching to take it off.

I clutched it around me.

—I like a big coat. I can roll up the sleeves. How much?

—Ten dollars, Gran said.

—Give it to him for five, Sylvia said.

—Ten is fine, I said.

—Five is fair, said Gran. Looks ridiculous.

I gave her five dollars.

—That's tax-deductible, Gran said. What about the car?

I had been hoping everyone had forgotten about the car. I just wanted to get out of there with my tax-deductible coat. Then I could think about things.

—I guess I'll wait on the car, I said. I don't really have anywhere to park it.

—Nowhere to park it. Park it anywhere. You said you wanted it.

—I said I wanted to *buy* it.

—A dollar.

—Gran, please keep your car, Sylvia pleaded.

—Getting rid of it one way or the other.

—The grandkids will be mad if you give your car away to a stranger.

—He's *buying* it. Those kids don't want this car.

—Gran, please!

—And he's going to mow the lawn.

—How about if I just borrow it, try it out?

—Borrow it, then, but I don't want it back.

—Gran you need your car.

—Buying a new one, I told you. Your uncle's taking me looking. No foreign cars, I told him.

Sylvia gave Kaiser a rough rub on the neck and stood up, and the dog, shocked back into the harsh uncertain world, squirmed backwards on its belly under the car.

—Fine, Sylvia said. Bryce, your new car.

—Where are the keys? I said.

—Doesn't need a key. Just don't lock it. Trunk opens if you bang on it.

She walked to the car and soundly thumped the trunk twice

with the heels of both hands. Up popped the lid. I threw the overcoat in, and slammed it closed.

Beware of the expedient action, I thought; but, I had been wanting a car, for shopping, for assignments that took me outside the neighborhood, to get out of the city once in a while. I really did like that old car.

—Thanks a lot, I said to Gran. I'll come back to take care of the yard, I guess.

—Better make it soon, it's all starting up again. You can take down those apple trees, too.

I nodded, surveying the thick twining growth, the gnarled trees. Wondered if I had just signed my life away: "While you're at it, till up the back forty and put the crop in, would you?" But there was no protesting further. This was a done thing, and again I had that odd sense of a predestined ending. Gran offered automotive advice as we said good-bye. Sylvia took my hand as she walked to her car, and as we stood by the door she said:

—Why don't you come to my house?

—All right, I said. I'll follow you there.

Gran had to pull Kaiser from under the car by the chain. The ignition, as promised, did not need a key—two little wings protruded from the fixture around the keyhole, and you just turned them with your hand and it started. The dog looked frantic as its haven rolled away down the drive. What if the sky should fall? I backed out past the gray apple trees: squirrels.

There were a few working farms left out here, but soon we were in the thick of the new wave of suburbia. Turned onto the freeway at the new mall. How strange it was to be driving this huge old car. I sat up straight on the broad bench seat, glittery blue fabric, nowhere a tear or stain. I felt like an old dad. I rested my arm out the window. The big tires rolled around with a comforting sound. Ahead of me, Sylvia's little car seemed to be going much faster, seemed to scurry, while I barreled along with this extraordinary momentum. Our minimal caravan rolled toward the city from the north. Downtown came into sight. The river cut in front of the downtown buildings; the locks and dams gave the river a look of purpose; the old flour mills and grain elevators fronted the steel-and-glass skyscrapers—the city's history in layers like geological strata.

Sylvia signaled for the University exit. I had to force the Bel Air to follow. That car, as it turned out, had been to church too many times, and now it yearned for the open road. It wanted to keep on going, south, through the night, across the rolling southern hills and on across Iowa, hellbent for Texas, better yet, Mexico. That was a car that, once it got going, didn't want to stop. But you can't just drive all night to Texas, can you? You can't let your car tell you what to do. I followed Sylvia onto the exit.

She pulled onto a sidestreet, stopped, got out of the car, walked back to where I had stopped, and bent to the window.

—My house is the third one in on the alley, the red house, she said. It's a women's rooming house, we're not supposed to have men in our rooms, so you'll have to sneak up the back way. Park here and go through the alley. Climb the fire escape all the way to the top. Duck down at the second-floor window.

I didn't have time to say anything, and she was gone. She pulled ahead in her car and turned into the alley. I shut off the Bel Air and followed on foot. I found the house, with Sylvia's car parked in the dirt lot behind it. It was a three-story Victorian ruin that fronted on the East Bank's main street. I saw the fire escape that snaked up to a small door, more like a hatchway, converted from a dormer window, with a narrow pane of glass on either side. It was getting dark now, and I dodged the square patches of light thrown out from the first-floor windows, and started up the fire escape. The steps were shaky, the round rails rusted. I must have been making a tremendous racket, I was sure the trembling vibration could be felt all through the house. I ducked down as I passed the tall arched window that gave on to the second-floor hallway, shrinking from the pale arc of light it cast. When I reached the top I was out of breath, a little dizzy. I peered in the narrow windows, could see nothing. They were the only windows in the room, and faced north. The door was painted red, the paint cracked and peeling. Sylvia is a hiding person, I thought.

I sat on the narrow platform outside the door, feeling hugely exposed. It was nearly full dark now, and I was glad for it. I sat and sat. Sylvia, where are you? The corrugated slats dug into my backside, my leg. I tried to sit perfectly still. As it became darker, and I sat tense and precarious outside Sylvia's door, I began to feel I was floating, or teetering on a high trapeze. I smoked a cigarette, and watched the ash fall down through the long darkness. Waited for Sylvia, a vertiginous anticipation.

Two bars of light sprang out around me. The sound of a door closing brought me back. The red hatch behind me opened inward, and I blinked into the light around Sylvia's shadow.

—Mrs Simic started chatting, Sylvia said. She wouldn't let me go. Sorry.

She stepped back and swept out her arm.

—Won't you please come in?

I rose to a crouch and turned stiffly, felt the pain return to my numb butt, felt the chill that had fallen with darkness. I ducked my head and pulled myself through the makeshift entry, the last in a day of strange doors.

Beautiful Swimmer

I woke looking at the ceiling that slanted over Sylvia's bed, at the lines of the rafters running dark through the yellowed paint. I looked with happy detachment: now there's something I haven't seen in the morning before. In the joyous bleary instant of waking all things are curious and benign. Sleep has lifted away your frame of memory and responsibility, and you're not really *you,* you don't have to be. You're like a happy rat. Waking in an unaccustomed place intensifies the sensation, and I wasn't even aware of Sylvia, at first, until the frame dropped down with a panicky jolt and with it the physical world, sudden and there, and the slanting wall was too close, and Sylvia's back pressed hard against my shoulder, pinning me in the narrow bed of which Sylvia had more than her share. But one of Sylvia's legs was thrown back and lay hot across mine, her foot brushing my ankle, and that was not a feeling that I was anxious to lose. I couldn't lie still, though, and I wriggled to free my arms, bumping my shoulder on the wall as I turned to the side. Sylvia shifted, moved away slightly. I lay propped on my elbow with barely enough clearance. I looked at Sylvia, so soft in sleep, her short dark hair tousled like a child's. I studied the freckles that patterned the skin of her shoulders. Her face was pink, a little puffy, her nose squashed against the pillow. Her pillow was wet where she had been lying—Sylvia drooled in her sleep. Then as I watched her she stirred, turned to me and raised her head, and looked at me with an expression of profound seriousness, as if she were about to impart to me the most important, most meaningful secret she harbored, something she hadn't even told Hermes; as if, in her sleep, the secret key to all existence had been given to her and she intended now to speak it—then her focus drifted, her eyelids dropped heavy, her head fell back to the pillow, and she was asleep. Probably she never woke up.

A pale ray of silver light somehow found its way through the north-facing windows and shone on the side of Sylvia's face. This light showed a soft nap of down that covered her face, her nape. And the crease beside her mouth, the divot from nose to upper lip, gave her face a feline cast. I shifted, and she moved and stretched with luxuriant ease.

I wanted her, but I was confused. If I pressed myself to her, kissed her neck, touched her breast, which Sylvia would I wake? Difficult, distant Sylvia? tender, giving Sylvia? or another? and how many Sylvias were there? Too many questions for so early in the morning; diffidence and doubt overcame desire. I maneuvered to arch over her, climbed onto the round braided rug. Stood there naked in Sylvia's tiny odd room.

Across from the bed there was an alcove in which Sylvia had

her desk—yard-sale issue, painted glossy brown, three side drawers, one white, one green, one missing. Three shelves above the desk were crammed in no good order with papers and books. A molecular model fashioned from red atomic knobs and blue straw bonds teetered on the edge of the lower shelf. There was a closet beside the alcove, its door covered with a bamboo-print curtain, blue and gray. On the hallway door, under a cracked plastic cover, was a sheet of paper listing the house rules, of which I had apprised myself last night. They included: **NO MEN; NO PETS; NO ALCOHOL (also no beer or wine); NO COOKING; NO SMOKING; NO MORE THAN THREE (3) SHOWERS PER WEEK; NO CHRISTMAS TREES.** Certainly naked men were harshly forbidden. I found my clothes and quietly dressed.

At the foot of the bed, beside the hallway door, stood a dresser, and above it hung a wood-framed mirror with photographs and postcards stuck between glass and frame. A half-sized refrigerator with a hot plate on top anchored the other side of the door. Desk and chair, bed, dresser, refrigerator—these were the complete furnishings.

Beside the hot plate were piled the dirty dishes from last night's dinner; and the air smelled of sautéed garlic and onions. And it smelled of cigarette smoke. And it smelled of the scent that had teased me the other morning through all the heady odors of spring's awakening. And now I knew that scent, now I was drenched in it; I smelled it in the fabric of my shirt as I pulled it on, I smelled it on my hands; I never wanted to lose it.

It was barely dawn, or so it appeared through the narrow windows that were dusty, smeared. The light that had touched her face had taken a difficult way. The right pane had a diagonal crack, sealed with peeling Scotch tape. Sylvia slept on beneath the angled wall. Good eater, good sleeper. Metaphysical Sylvia did not neglect her animal needs. This I contemplated fondly.

I sat in the chair at Sylvia's desk. The books on her shelves were linear algebra, biochemistry, physics, poetry. I took the molecular model down, flexed it gently in my hands. It was fairly big, something complex. It was just wood and plastic, but I felt a kind of energy in its architecture. The blue bonds shot out this way and that, angled to bind the dozen or so red atoms. It sort of pleased me that I saw straws and knobs, while Sylvia could look at it and see periodic symbols, atomic weights and numbers, specific gravities, chemical properties. Shared electrons buzzed between the atoms in a physicist's cloud of unknowing; in the nucleus protons and neutrons held like anchorites. Engine of the universe: straws and knobs. And maybe this structure wasn't anything specific at all. Maybe Sylvia had been creating matter, just to kill time.

Again I thought of Carla—a twinge of betrayal. The Sylvia she loved, maybe, was that Sylvia in chemistry class. But Carla's maybe was growing small in the distance of the yes that Sylvia and I had said last night.

Sylvia shifted, rustled and kicked, throwing back the covers and exposing one leg to the hip. I was dizzied by this sight, and by memory: how we had touched: this leg of Sylvia which she had given me to caress, the skin of Sylvia: the wondrous skin of Sylvia which shaped to her so intricately, so splendidly, which wrapped the form of Sylvia whom, maybe, I loved. It was astonishing, how we had lain entwined, and not demurred from touching every inch. Could it ever be understood, except in its enacting? Now I didn't know if I dared to touch her.

My bladder insisted relief, but the bathroom was down the hall and I didn't want to risk being seen. I wanted water to clear my gummy mouth, and found a dusty inch in a glass beside the bed. I lit a cigarette and sat smoking, and felt oddly self-conscious. A bare bulb on a tattered cord (great cliché of rented rooms) kinked down from the splotched ceiling. Something about its placement bothered me—it hung just slightly off center, if the room with its angles and slanting walls had a center. Watching Sylvia sleep, I felt like a voyeur. The room felt intensely personal, deeply *Sylvian,* though it was not really decorated at all—the only things on the walls were the mirror and a cheap print of "The Last Supper" in a cheap frame over the hot plate. As if Sylvia had lived here all her life. As if no one else had ever set foot here.

It was all too much, and so I left.

The little hatchway door had to be locked from the inside, so it stood open a couple of inches when I went out. That would help clear the cigarette smoke, give Sylvia good morning air to breathe. It was cool and damp out, a light mist softening the distances. I came out of the alley thinking it was too early to hope for a bus, but not minding the idea of walking home, and turning the corner came head to headlights with Gran's Bel Air, waiting there like the blue car from hell. *My* Bel Air. Then not the car from hell, but my trusty old conveyance. Hello, Old Girl, was it cold in the night? I climbed in on the springy front seat that made you want to bounce up and down. Turned the ignition and she started first time. Now the day was going good. It was ridiculous. I wanted to drive.

But I had work for the morning, and so had to settle for taking the long way home. I followed the river from the southeast city to northeast, past the flour mills, the locks and dams that allowed passage over the falls (the reason the mills were here, the reason for the city; thus are our lives ruled by geology—why do you live where you live, you in your deepwater ports, safe harbors, deltas, and confluences, in your richly silted valleys, last-stand foothills? Rock and water are everywhere the cause). From northeast I crossed to downtown, and rolled along the broad main way, empty at this hour, surveying the buildings with a proprietary air, like a riverboat pilot recognizing islands, bars, and crags—this car rode like a boat, after all. The basilica at the edge of downtown aspired to a copper-green peak, with a weak sheen under the misty sun. I turned there, and then again, onto the parkway that carried me into the wealthy green neighborhood where I had grown up. The car mastered hills and curves with splendid ease—the suspension was fine. The AM radio had that wonderful tinny sound you just can't find anymore, buzzing out of the metal dashboard, and I almost expected it to capture some vagabond signal that would bring back the beloved deejays of my youth.

The parkway brought me down a long hill to the lakeside boulevard. The water was still, and nearly the same green as the basilica dome. Around the first curve on the boulevard I came upon my mother's house, my childhood home. She lived there alone now, since Colin, my older brother, and I had moved out—Colin purportedly to graduate school in the East, shortly after our father's death; I to a university dorm about the same time, with occasional extended visits, and then to my apartment in a cheaper neighborhood, not far way, about a year ago.

I smiled as I drove very slowly past the brick house, two-story, modest for this neighborhood where some of the houses had elevators, ballrooms, and areas identified by separate names other than "upstairs" and "downstairs." My mother might have been looking out the window, seen this old blue boat cruising by, and the identity of its driver would be a mystery. I thought about stopping, honking, driving on. She still wouldn't guess. Having this car was like having a new identity, driving it gave one an entirely new perspective. I hardly felt in control as we wove along the one-way boulevard. It was like trying to drive fate. The car felt *inexorable.* I let it run a stop sign, because the boulevard was empty, and it seemed to really want to.

I drove around all the lakes—there were four in a row on the city's western edge, puddles in the path of a glacial river. By now the streets were busy, noisy, and the spell of solitary seeing was broken, and I took the straight road home.

Yes: unspoken, unknowing. In the morning you remember what you said in the night. Do you regret it? Did you mean it? Do you understand it? If we belonged to some people that knew itself a people, what we did might mean that we were wed. By the choice we made. Where acts have meaning more sure than words. We take this act, all acts, all words, more slovenly; but can we pretend it means nothing?

I made coffee and sat at the window. I considered the day ahead, but schedules, assignments, and deadlines held little interest. I felt a little sick, a little dizzy, almost on the verge of tears, but just as likely to burst out laughing. What joy is joy?

There was a knock on the door. Was it my neighbor, coming for *Two Gentlemen of Verona?* I still hadn't read the whole story of Silvia. But it didn't matter—if it had been a sign to guide me, it had done its job. I went to the door and opened it on Frank. A relief. I let him in, offered him coffee.

—No, no coffee, he said. I'm going for a swim. Come along.

I remembered the cold look of the lakes under the mist. I shivered.

—Are you out of your mind? The ice is barely off.

—It's a perfect morning for the first swim of the year, your baptism, Frank said. You look a bit ragged. A swim will wake you up.

He went on trying to convince me, because he knew he could. This morning I thought I was adamant, but as I turned from the stove where I had been pouring coffee, and really noticed Frank for the first time, leaning against the refrigerator dressed in a gray sweatshirt and white athletic shorts, his long pale legs descending to a pair of red beach thongs, towel thrown sportily around his neck, his brown eyes bright and mustache bristling cheerily, his dark curly hair curling, I was lost.

—I'll come with you, I said. But I'm not going in.

—Fair enough. Grab your suit and a towel, though, just in case.

We went out the front way, and started to walk up the avenue. I stopped.

—Wait! I said. I have a car!

I drove to the northernmost of the city's lakes. Frank was delighted with my car, and didn't even ask where I'd gotten it. We rode along smiling. The lot by Frank's preferred beach was empty. It was far too early in the year for swimming, any sensible person would say. Frank immediately stripped down to his tiny blue suit and went wading straight in from the small beach on a cutaway bank. The lake was surrounded by trees, with houses showing here and there through them. Docks stood empty, swimming floats quiet along the opposite shore. Farther down that shore the lake dipped into weedy bays. Frank waded to waist-deep water, then extended his long arms over his head, hands together, prayer-like; he arched forward and disappeared beneath the still surface. I watched, enthralled, for him to surface, as if he were Houdini, locked in a box. What a gift of beauty and artless drama as he ascended to the air, arms still extended, legs straight and rigid, as if that one push had carried him all the way. With only the slightest pause, a miracle of timing, he spread his arms in a great sweeping stroke, coiled his legs and released, and in no time was far out in the lake. If he stopped swimming would he simply float? He swam that effortlessly. This lake wasn't big enough for him, he did circles over the deepest deeps. His slick brown head bobbed up like a seal's. He was not graceless on land—slightly awkward, with his long knobby limbs—but in the water he was another creature entirely, water-baby, changeling, never fully himself in our dusty life. But, enough of himself anywhere, I thought.

Frank was a dark head darting, a smooth dying wake, far out in the water. I had brought my suit, and watching Frank had inspired me. I stepped behind a tree, stripped, pulled on my suit, walked tenderly to the water, dipped in a toe. I almost gave it up at that, but some foolish bravado impelled me to continue. I waded out a few unsteady steps, gooseflesh rising everywhere. At mid-thigh I realized I couldn't be any more miserable, and so I plunged, came up gasping, thrashed out a few strokes, made a panicky turn and headed for shore. Came onto the beach floundering, flinging water and sand. I used my towel, and then Frank's, because I was not so enchanted with him now. Behind the tree again I peeled off my clammy suit, dressed, added Frank's sweatshirt on top. I sat on a rock, shivering, and smoked.

I considered driving off and leaving Frank there with a damp towel and no shirt (what power and freedom in this automotivated existence! you could just *drive away!*). But it wasn't Frank's fault that he was a beautiful swimmer and I a poor one, or that the water was cold. I recovered enough to realize this. And as I slowly warmed up there was a nice tingle to my skin, the lake was pretty; I didn't mind waiting. I was indeed awake, and thinking fairly clearly.

What else was there to think about? But why was she so hard to hold in mind?

Isn't that always the way with the other-worldly. She hardly seemed real, she seemed almost mythic and somehow tragic. She had suffered torments in the cedar closet because of the accident of blood; familiar of saints and bronze gods; initiate to strange arts. When I recalled simple human details about her, they were almost shocking: how her sock cuff had rippled her pale calf; how she drooled in her sleep; her uncertain expression as she stood alone at the party. She felt so distant as I thought of her, it was hard to imagine that we had kissed, that we had made love. Yet I remembered. To live like a shrew, would she give up that memory?

When I ducked through the small door into her room it was as if I were being ushered into some wondrous secret chamber, a splendid cell at the heart of an intricate labyrinth. It struck me as a cell, a hermetic keep, if you will. And this pleased me. She made us food, spaghetti with red sauce, apologizing that it was not up to Carla's standards—she mentioned the name easily. She brought out a bottle of wine. I drank wine and smoked as I watched her cook. I noted the list of rules on the door—four broken in that moment, flagrant disregard, and I was delighted. The reflections in the small windows turned the room inward. Had I ever felt so safe, so privileged?

Sylvia ate with excellent appetite. We had no problem finishing the wine. She apologized again for snapping at me at Gran's; she said it was all old business, and to pay it no mind. We sat on the bed, the only soft seat in the room. She drew up to me and we hugged, held each other chastely for some time, and then kissed, and kissed and touched. And then she rose, and went into the closet. I thought she might disappear through there, as if the bamboo curtain were yet another magic portal. She returned with a diaphragm, set it on the floor within reach. She said,

—It's a little rusty—not literally. *I'm* a little rusty—not literally. I hope it's still okay.

Bodies do what bodies know, what the mind will lead us to without knowing why. Is there a physics of desire? As for intentions, plans, decisions, I think we do what we will, and later we tell ourselves little fairy tales of why.

Now I saw her rising and finding me gone. Was she angry? Did she feel used? No. She smiles. She is feeling happy and slightly silly. She steps out of bed and stands naked at the cracked window, wipes sleep from her eyes as she considers the day. There's a draft and she shivers and pulls her arms around her, sees the door tilting open, and reaches over to latch it, but she's hardly awake and it takes a couple of fumbling tries before she gets it. This effort has distracted her, she stands a moment without a thought. She takes a halting step forward, toward the desk, but that's not where she meant to go, and she turns abruptly, reaches into the closet through the curtain for her robe, a towel off a hook in there, the dark contours so familiar she doesn't have to look, and with one hand holding closed the robe, the other carrying the towel, she goes out the door and down the hall to the bathroom.

It's chilly in there, because Mrs Simic is cheap with heat, and Sylvia turns on the shower hot to steam the room while she pees, brushes her teeth. She looks at herself as the mirror begins to fog, and yes, it's you again, Sylvia, this morning as every other. Maybe she's surprised that she continues to exist in this solid way, day after day, and wonders if the face that quickly blurs is really the shape of what she is, if she is what people see; maybe Sylvia has a little trouble imagining her, too.

She adjusts the shower and steps awkwardly over the edge of the tall claw-footed tub, and now the sheer physical joy of body and water, of being warm and wet, washes away all metaphysics. She sighs and gives a happy groan as the water pummels her face. When she's washing and a little awake, she thinks of what she has to do today, of finals coming up, of the project ongoing at the lab; she has a schedule in her head, every duty in its place. Thinking of finals, her pulse quickens, but she has every class aced, regardless.

Then as she rinses soap from her hair, she remembers last night, and her heart jumps now, too, but in a different way from when she thought of finals.

What way did her heart jump?

One mustn't presume too far.

I wandered along the shore as Frank glided under a brighter sun that burned through the mist. I could completely forget he was out there. He could have navigated through the channel to the next lake down, or drowned, I wouldn't even have noticed. Sometimes being too good at something is dangerous.

I looked at the green green leaves, so new and whole and intricate, but looking closer I saw that they were already galled and eaten, almost every one marred somehow. Frank startled me, surging up on the beach like a pale emanation. He shook his head and water flew and his curls rebounded. He rubbed his face, preened his wet mustache.

—Wasn't that good? he said.

—Good for you, I guess.

Frank stood wet at the water's edge. He turned, tilted his head, surveying his surroundings as if he had just come ashore in a new-found land. Everything about him was long—neck, arms, torso, legs. His chest was smooth, a little sunken, his hips narrow and bony. He came toward me where I sat on the bank.

—Life is good, he said. Are these not the very sands of paradise?

He swept out his hands to indicate said sands. I in turn presented the apparently untouched shrubberies, and said:

—Every leaf is eaten. Barely born, already et.

—Well, said Frank. It happens.

This stupidly accepting optimism was something I loved about Frank. Maybe because I didn't share it. I tried to be optimistic, but it was like the irony thing: at heart I was sincerely pessimistic. A line I had read in a Poe story had become the keystone of my philosophy, and I repeated it now to Frank:

—It is written, that in every man there is a hidden quality, the antagonist of bliss. Do you believe that?

—Who said it?

—What difference does it make who said it? Just answer yes or no: do you agree?

—Makes all the difference in the world who said it.

—Poe said it.

—Well there was one sad character. No wonder.

—What if a happy person said it, say Ronald Reagan?

—Ronald Reagan wouldn't say that.

—But if he did, would you believe it?

—If Ronald Reagan said it, the world would be catastrophically different from how I thought it was, so I wouldn't know what to think. But why worry about what depressed writers or deranged presidents say? If my grandmother had wheels she'd be a trolley bus. If wishes was fishes we'd have a big fry. This good morning we tread the sands of paradise. Enjoy.

—You're so shallow.

Frank laughed as he took his towel, and he noticed that it was damp but didn't mention it.

—To be deep you have to suffer, he said. Who needs it?

At the moment my concentration had shifted from our conversation to the fear that Frank would soon ask for his sweatshirt back. I hugged my arms around me, as if he might try to regain it by force. He went on:

—But you can't really avoid suffering, so you do what you can, try to control your own little karmic sphere. The rest is quite beyond you.

I realized my best chance for keeping the sweatshirt was to keep Frank talking, so I said:

—Why are you always telling me things I already know, as if they're revelations?

—They are revelations. I reveal to you what you already know.

—Why not tell me something I don't know, just for variety?

—You wouldn't believe me. And, it probably wouldn't be true. Like Reagan turning into Poe.

—I knew you'd say that.

—There you are. All our efforts, and our basic survival instincts, are turned toward confirming the world we believe to exist. Anything that doesn't fit, we discount or disbelieve. The narrower your conception of life, the greater your chance for happiness.

—Uh-huh, uh-huh, I said, which was what I usually said when Frank started going off.

I believed Poe, and I believed that Frank deep-down was as miserable as anyone, and that his blithe exterior was the product of a compelling illusion. Frank was a painter and knew how to make things seem as they were not, but I was a mere reporter, and had to contend with things as they were. Which, as I say, I was not very good at. Frank stopped talking as he brushed the grit of paradise from his feet and dried between his toes.

—I'm sort of seeing this friend of Carla's. We met at your party.

—I'm so pleased, said Frank.

—That's her grandmother's car I'm driving. Her grandmother gave it to me.

—How generous.

—But I have to mow her lawn.

—Sounds fair.

Frank pulled on his shorts and stepped into his thongs, and sat beside me.

—So it's love, then, that's making you cranky, he said.

A pair of joggers, man and woman, ran by on the paved path that circled the lake. They were trim and pretty, in brightly colored shoes and shorts and tops. They scampered out of sight into the lakeside greenery like little animals.

—Do you think those people were in love? I asked Frank.

—Yes, he said.

They didn't wreathe their arms about, sigh or weep; they appeared anything but pestilent. That wasn't the modern mode of love.

—Do you think that love conquers all? I said.

—No.

—Why not?

—If love conquered all, it would have to conquer love itself, and can the conqueror also be the conquered?

All out of the blue I remembered Sylvia's poem, about imaginary numbers. The imaginary number, identified as *i*, represented the square root of negative one. But negative numbers couldn't have square roots. It didn't make sense, but in the calculus of imaginaries, it did.

—You're pretty logical for an artist, I said.

—I used to be logical. I remember that from my logical days. Now I try to live intuitively.

I reached for Frank's hands and we pulled each other up. Turning toward the parking lot, we were almost run down by another pair of joggers, equally trim and bright as the first pair, and perhaps somewhat prettier, but as these two fled into the greenery it seemed to me that they were not in love, and that this, in spite of all logical complaints, made all the difference. Live intuitively; the mind founders, but the imagination plunges on, like a beautiful swimmer.

Smart & Pretty

Rinsing soap from her hair, Sylvia thought of last night, and her heart jumped again, but in a different way from when she thought of finals. Whatever it was that made her heart jump this way then rose up through her chest, her throat, with a prickling warmth that exceeded that of the hot water, and Sylvia laughed, took in a little dilute shampoo, spit it out, laughed again. She could not remember being this happy in the morning.

The exhilaration remained as she stepped out of the tub, toweled off in the steamy bathroom. A gentle tap at the door—one of her housemates wanted in. Sylvia felt a little guilty, knowing that she had used most of the hot water, but this did not spoil her mood. Combing her wet hair she thought of finals again, and now her heart was jumping two ways at once. Sylvia paid attention to her two-ways-jumping heart. She knew that whatever you felt meant something. She knew that intellect was compact of heart and mind. If her heart had jumped more often in the second way it might have made her the poet she longed to be instead of the incipient scientist she was, though she still had hopes that she could one day be both. What one became, to Sylvia's mind, was that which one could not help becoming.

Some things must be said now about Sylvia, about what she had been, and what she now was, and what she was becoming. These are not easy things to say, there is much at risk.

First, Sylvia was, had been, always would be a remarkable student. She was intelligent, and could have earned good grades with little effort, but that was not enough; irrefutable excellence was all that would satisfy her, and not even that, and it cost her great effort and anxiety. She would settle for nothing less than an A,an A-plus if possible, and Sylvia made scant allowance for the impossible. Hard-nosed veteran teachers, appalled by grade inflation, who had never granted an A-plus and vowed each night as they went to sleep that they never would, remembered Sylvia as the tiny fierce girl who had wrung one out of them, and they woke in the night, years later, thinking about her. What teacher didn't dream of having, just once, a student like Sylvia? What college admissions officer didn't lust after her: her transcript with its glorious alphic monotony; her extracurriculars—newspaper, yearbook, choir, literary magazine, student council; her prizes—German, English, Math, Science, and sundry for all-round excellence; her SAT scores, 1520 combined? Sylvia was good in school.

Sylvia acquiesced to systems. She never worried that she was merely fulfilling the expectations of others. She seemed to understand very early that a force requires a medium, and if others, for the time being, provided that medium, she did not feel her force diminished by it, her creativity thwarted. How many great paintings have been summoned by commission?

Among her fellow students, then, Sylvia seemed charmed. But who knew her torment? Who knew how she suffered if, near term's end, she somehow found herself teetering toward an A-minus, or, dear god, a B-plus? Who knew, and indeed, who cared? No one cared. Get a B, Sylvia, they would say. Loosen up, it'll be good for you. And it would have. But she couldn't.

The next thing about Sylvia is this: she was pretty. This has been stated previously, and unfortunately it must be repeated. It wouldn't be fair to make Sylvia homely just because she is smart. Still, the danger in this is clear. She was not drop-dead beautiful—and oh yes, she was short, and her legs were stout though not unattractive, and her skin tended to be a little dry. She was not an impeccable dresser. Just the same, when the boys in high school and later in college considered whom they might like to ask out, Sylvia was not at the top of the list, but she did come to mind. Boys liked her and Sylvia liked boys, though she scared a lot of them, by being too smart. A boy with any intelligence doesn't like a girl who's smarter than he is. In rare cases an enlightened boy will allow that his girl is more gifted than he in some ways, but he always harbors a secret superiority. Some men overcome this, or manage to pretend that they have. Women on the other hand rest certain in the knowledge that this question doesn't matter at all; this disinterest they consider simple charity, to spare men the embarrassment of their lacking should accounts every truly be totaled. But not many of even the bluffest of boys cared to stand comparison to Sylvia, so though she was admired by many she was pursued by few.

It did happen, though, that in her last two years of high school Sylvia had a grand romance with a boy who was confident, handsome, smart enough, and a year older than she. His name was Chad and his family had money. Well, he wasn't that smart. He seemed smart enough because Sylvia really loved him, or the part of her that really wanted to love someone and have someone for her own really loved him. It must be put this way because Sylvia had a kind of two-sidedness to her—the disciplined, demanding side that drove her in all intellectual endeavors, and the side that despised the first side and wanted nothing but total abandon, abnegation of the self. It's convenient to be able to put things in words like this, draw attractive ideas like this out of nothing but what happens helter-skelter and then is gone. Just looking at Sylvia, you might not have known this about her. But one's past does tend to become schematized in this way, and anyway, what's the point in going into great detail over a teenage romance? We deal with the present in great detail, and with the past in its larger outlines. The past is there in the present, its symbols ever changing.

With Chad Sylvia indulged the side of abandon. Indulged as if she were trying to deprive the other side of any advantage. As if Sylvia were saying to Sylvia: there, you've been out drinking all night, haven't slept all weekend, let's see how you handle advanced-placement German, now. One wonders how Sylvia felt when Sylvia handled it just fine. Perhaps in moments of bleary blending the two sides could call a truce, share a brief appreciation of their respective accomplishments.

Sylvia and Chad continued to see each other when he moved on to the University, and Sylvia was saved from high school ennui by the bland exoticism of the fraternity boy's world. This time held for Sylvia a gilded aura. It wasn't beer and puking in the bushes and stupid hats and sports that she recalled. Her mother had taught her to Lindy and Charleston, and one night she made Chad take her dancing at a club downtown. Some of the other boys and their girls tagged along. Awkward, out of place, they didn't stay long. But then in memory, every night had been Lindy-till-you-drop. She made Chad call her Zelda. She was out after madness, or so she conceived of it. Sylvia's mother disapproved entirely, and for Sylvia this was to no small degree the point. It almost came to Sylvia's moving out. Sylvia's parents had been long divorced now. Sylvia's stepfather, Richard, refused to involve himself, and wisely. Her father, who lived in a kind of self-imposed exile in a suburban condominium complex nearby, knew little of it, except that Sylvia and her mother weren't getting along, which was not news.

The lover descends: What bird / this time, and what new world? Sylvia cultivated madness for the sake of poetry, which was unfortunate. *He comes forth from the mazy / tunnels of intellect to claim her / in the only way. / The beloved responds: / I could not but acquiesce.* Sometimes it seemed that Sylvia understood everything, but this she did not yet grasp: that science is passion and poetry precision. Byron came home and conscientiously composed after many a grand debauch, but Sylvia imagined that the poetry was in the midst of it. *Lover, North Star, faithless guide.* Her two sides tried in turn to enlighten her, but there was a third element involved, which was Sylvia's will, and nothing trifling. Perhaps such prodigiousness requires such a flaw as errant will.

The last thing about Sylvia, for now, is this: she had suffered a fall. Toward the end of her senior year, with colleges courting and a great prospect broadening, Sylvia tired of Chad.

She was growing up a little, and he didn't seem to be, and Sylvia finally had to admit that he wasn't smart enough, that she didn't love him as she had liked to imagine in her Zelda-esque extravagance of emotion. The sad part is that Chad really did love Sylvia, though she didn't deserve it. He truly did love Sylvia, because there was more depth to him than Sylvia at last was able to see, and he was left heartbroken and bewildered when Sylvia abruptly ended it. It would have happened soon enough, regardless, but Sylvia did not behave well in this.

Has it been mentioned that Chad was Sylvia's first lover? After a fraternity party, after a football game, on an autumn morning just around dawn. They had been going together just short of a year.

Sylvia had her pick of colleges. Money was a problem. She and her mother fought over it. What was wrong with the University? Irene had gone there. Sylvia's father had gone there. Sylvia's father couldn't help as much as Sylvia thought he should. Richard tried to introduce compromise, conciliation, but Sylvia, her mother, and her father were going around in circles too fast for him to enter.

It was somehow worked out, with a few hundred dollars left to chance, and Sylvia went off to her prestigious, expensive university in the West with really no one's blessing. Sylvia was headstrong, somewhat arrogant, unbending, but please! She was only seventeen! That was no kind of send-off. And so we can't blame her entirely for what happened next.

What happened next: First, Sylvia was not the only bright kid at school. All the kids at this college were bright. College is harder than high school. That hardly needs saying. But it wasn't that Sylvia couldn't do the work, there was really no question about that, either. Still, it required more of her, and she had let her skills of concentration slip in the last year, becoming valedictorian by the seat of her pants. And she was far from home and entirely alone, as she had thought she wanted to be,but now she had to wonder why she had. Another poetic fallacy, romantic isolation as a triumph over circumstance. Sylvia had no friends. That was her fault. Now she didn't know how to make friends.

Things piled up: Sylvia's roommate was even smarter than Sylvia, and more closely acquainted with madness. Early in the quarter her paranoid delusions began to surface: her professors gave her low grades just for spite; the dorm R.A. read her mail; someone was sprinkling LSD on the salad bar. Sylvia came home one day to find her stuffing wet sheets and towels around the windows and under the door—gas! Someone was trying to gas them. The girl went home, or somewhere, midway through the quarter. Sylvia, alone in that room of protean terrors, rather missed her. She had some acquaintance with senseless persecution; the specter of the *someone* reached her deeply. Et tu, Zelda? She thought less fondly of madness now. All that fall she did not dance.

She went to class, worked as a set-up/clean-up assistant in a biology lab, studied at the library until it closed, came home and fell on her mattress exhausted. With increasing frequency, without warning, she wept, and found that she couldn't keep herself, as she drifted toward sleep, from sniffing the air for a telltale scent. Then there was the money, and the strain of knowing that no one in her family wanted her to be where she was. Sylvia had tried to believe that she had no family, that she was an orphan, post priori, or a self-created being, sprung full-grown from her own head (maybe Hermes had some part in that). But one's family, even if it is splintered and at odds, does generate this field of force. Even her sisters—Rachel and Alison, three and eight years younger—were against her, not because they cared at all what their weird sister did, but because of the general disruption that Sylvia had caused. Whenever

Irene and Sylvia were together, whenever Irene merely thought of Sylvia, she became more arbitrary, capricious, and ill-tempered toward everyone. Patience and tolerance were not bred deeply anywhere in that family.

By the end of the first term Sylvia was through. She had to admit it, she couldn't stay there anymore. And Sylvia did not like to quit. There was a measure of satisfaction in deciding to go home, and much regret. She had found a limit, and in time that would strengthen her. Not right away. She finished her courses in good Sylvia fashion (so the traces she left, though fleeting, were perfect), informed the dean that she was withdrawing, and flew home at winter break knowing that she would not go back. Only she and the dean knew this.

But, home . . . This is difficult. She flew in on a clear cold night. What thoughts and emotions riddled Sylvia as her city came in sight? The drowning love and longing of a first return. This place is so familiar, you bank and arc on icy wings above something almost yourself. The engines' straining throb is almost your heart, for you are undone and everywhere strewn by this thrill and sadness and desire. You are the sharp deep sound that thunders on the rooftops as you descend. And this is somewhere you have never been before. Down through the dark and dream-clear air, Sylvia saw bright blotches of city parks, and the immaculate ice surfaces of hockey rinks, rounded rectangles, marked each alike with red and blue lines and circles, which Sylvia abstracted to sparkling arenas of possibility; but they were small and widely spread, and mostly below there was darkness.

went easily at home. Sylvia needed her family's shel-
ce, but of course she would never let that be
had she come home? Because of loneliness and

spastic weeping and the fear of fearing gas leaks? No. Because the school wasn't what she had wanted, classes were too big, professors distant, wrapped up in research, her work-study job beneath her. It wasn't worth it, she could do just as well elsewhere for a lot less tuition. But her family had made these sacrifices to send her, and she owed it to them to stick it out at least through the year. Everyone owed something, it seemed, and no one was paying.

Sylvia enrolled at the University for winter quarter. After only a few days at her mother's house it was obvious she couldn't stay there. With Sylvia and Irene under one roof, every room burned with tension. Sylvia looked for a room and took the first reasonable one, that bleak third-floor cell in Mrs Simic's women's rooming house. She moved in the day after Christmas.

Through her first days and weeks there she felt disoriented. An odd homesickness plagued her. How was it possible that here in the heart of her own city she felt more distant and removed than ever? If you longed to be in a certain place, and then you arrived there, but the longing did not ease, then what could you conclude? That this place was not the longed-for place; that the longing was not for the place. Sylvia hated paradoxes. Now she lived in one.

She couldn't believe this wasn't the place she longed for. The winter city was to her almost intolerably beautiful. On some days, walking to class across the bridge over the railroad tracks that crossed the river to downtown, she thought her brain would burst with the beauty of the city, so sharply composed, as if through a lens, in the pure cold air against the richest blue sky. This was the place, then, but somehow she wasn't there, and in her mind she began to recede still further, and willingly, if the only way she could possess this place was in the strength of her desire for it. She cut out pictures from the newspaper and magazines—the stilled drama of the first big snow; mist rising off the winter river; summer thunderheads tumbling in from the prairie, reflected in the glass buildings downtown—and she tacked them to her walls, and lay in bed and looked at them, and thought what a charmed and lovely place this was, and how happy she could be if she were there. Sometimes she imagined that she was cutting out the pictures to send to a homesick friend who had moved far away: see how beautiful it is here, I know how you must miss it. But there was no friend, no far away, no here. Her tiny room turned in space.

One night in her room, Sylvia remembered, or realized, with a genuine panicked shock, that Chad's fraternity house was only blocks away. She was terrified by all she had forgotten, she felt she was losing her life. Chad and the dancing and the parties and her triumphs and her glorious future and her beloved city. What had happened? She knew. She had thrown it all away and that was why it wasn't here now. She had gone away one way, and she couldn't come back the same way, so she had arrived at a different place from where she had started from.

But maybe that wasn't true. Maybe not everything was gone. Sylvia had always trusted her conclusions, but everything had always gone well before. Now nothing was going well, and she was forced to consider areas of uncertainty. She stood at her window looking out at the bleak winter scene, bright under moonlight—gray rooftops, scrabbling tree branches. But there was life out there, hidden and seemingly distant, but there. Chad was out there, just blocks away. Some resolve was rising. Maybe Chad still loved her—surely he did—and would have her back, all she had to do was knock on that door. It would be a different life from what she had envisaged, but perhaps it was time for compromise. She could take a different way, trick paradox with a shortcut back to the place she had left from, be a different Sylvia, be humble and good.

She was almost frantic with remembering the warmth and cheer of the frat house (she had never liked it much) as she pulled on her boots and parka. She could feel the clenching dread flowing away in Chad's embrace (she hadn't even thought about him for months). She went out by the fire escape because she was about to fool fate. The streets were empty, the snow creaked loudly, her breath came out hard and white. She laughed with embarrassment when she had to think to recall where the fraternity was, and her laugh echoed off the cold houses. She imagined telling this to Chad as part of her penance. It took so long to walk there, she was half running.

She saw the house, dark stone, massive and medieval. Arched leaded windows, bright. The Greek emblazoning. The steps and walk cleanly shoveled by the dutiful brothers. They tossed snow in each other's hair, and wrestled in the snowbanks. They loved life. They drank and danced and loved life. They were strong and clean and they knew who they were and where they belonged, and if they didn't, someone told them. Their brothers. They told each other. A happy people in a happy world. Their girls were just like them, their sisters and lovers. The house was a hive and a haven, where you only had to be and that was more than enough, and the given love thrived in the very air. Music. A happy cook to feed them. Who they were. Where they belonged.

Sylvia reached the corner where the fraternity house created space with the power of an ancient temple. And the oracle spoke and said: Not you, Sylvia, not here, too late—turn away.

She turned away. She went to the river. She cried a little, hot and quickly chill on her cheeks, and then no more. She walked wrapped in silence and cold. She crossed the river to downtown, not pausing, contemplating nothing. Crossed back to the East Bank, and again to the West. It was bitterly cold, and very late; she did not think of time. She walked all the bridges and came back to the East Bank, climbed the fire escape, fell blankly into bed, and slept, and remembered no dreams. It was the first of many times that Sylvia would go out at night to walk the river bridges, forth and back and always back, as if looking for something that, even if it were there, she knew she wouldn't find, because she wasn't there.

Part Two

My Cheese

That bridge again. I hardly knew how I'd gotten there. The whole morning I'd hardly had a thought. I had dreaded getting up so early, before dawn, to follow the old soldiers to the cemetery on the river bluffs south of the city. But once I had risen, in the fresh quiet of late-spring dawn, an unaccustomed rhythm had captured me and carried me along.

Memorial Day.

The freeway had been empty, and a most promising road. At the cemetery the dark dewy grass lay flawless under the white stones, like an enormous comforter immaculately arranged. Turning my head I saw the little city, perched on the prairie like something lovely and distant, compact, intricate, desirable, attainable. I do not wake alert, and so each picture simply flashed its unclouded sense on the screen of my brain, and then vanished, and as its afterimage faded another flashed, so the images and sensations blended and lapped, blotches of color shading together to a seamless abstraction. And so with the images I discovered in my camera's viewfinder: a raised foot in heavy shoe showing its scuffed sole, with flopping trouser cuff stilled, and leading to it dark footsteps in the jeweled grass; a circle of old men bearing flowers, crouched in the delirium of stones, while airplanes in low approach to the runways, landing lights blazing, hung over the men like lamps, or angels. The early light somehow ravaged and redeemed the old faces at once. The viewfinder not only stilled its images but silenced the roar of the descending planes, so that, taking my eye away from the camera, I was startled at the sudden thunder. I shot many more pictures than I needed, because every time I looked I saw a perfect image. I knew that the perfection existed only in my eye, my mind, the moment, and that in print the images would pale miserably, the sharp contrasts and the splendid light all lost.

As I drove back to the city there were a few more cars on the freeway, but it was still early and quiet, and I felt a kind of kinship with those other drivers, and my mind traveled off in many different directions with many early travelers: three men in crumpled hats in a car pulling a fishing boat toward some mythical lake where trophies rose on the line like hidden wishes; an elderly couple, so neatly groomed, bound for a holiday gathering with the happiest family on earth, generation after generation in perfect accord through the certain knowledge of blood; a young couple with a cooler in the back seat, off early for a day on the sunny beach, or in some dappled wood where they might discover once more, or for the first time, the finest fruits of youth and love. Back in the neighborhood I shot a few more pictures of the veterans securing flags at intervals along the quiet main avenue, and I could imagine that the circle of the neighborhood ×-ed by the central streets had been lifted from the city and settled in the lee of still fields and misted hills. The people, when they emerged from their grimy apartments which had been transmogrified into tidy white bungalows, would have recovered the rituals and customs that bound them, and marched happily toward a future which led them at an accommodating pace toward a destiny they need not, did not, fear.

The dawn had said you can create or redeem a lifetime in a day, that there is time to know your heart, to reconcile and accommodate all will to world, so that the full light, when it comes, will show you nothing you are not prepared to see and accept. Now as I stood in the blank full sun of the aging morning at the West Bank end of the Washington Avenue bridge, I wondered where that light of promise had fled. I looked across the river, and felt as empty as the pale eastern sky. In the harsh light of everyday the fishermen rocked on the seasick chop of fishless waters, the family gathering had soured in the acid of every old grudge, whoever packed the cooler had forgotten the beer, and a winter's worth of litter in cyclones of dust swirled through the heart of Red Earth, more brightly than the manically flapping flags.

For me, that bridge again. Some searching impulse had gotten me this far, to the vicinity of Sylvia and memories of my father. I had trusted that the cadence of the day would announce the next measure in proper time. Hadn't I noticed the shrinking shadows, the hardening light, the air going stale and dry? As I stood on the wide concrete plaza I felt uneasy, exposed. The newer West Bank campus of the University lay off to my right, modern buildings of three or four stories squared around a mall broken by recent plantings. Since it was a holiday the mall was deserted. A few abandoned bicycles locked to trees or lampposts made me think of dissenters chained to embassy gates. Across the river was the old campus, its brick buildings obscured by trees. The wind had come up strong, as if to blow away the last tatters of morning hope.

Surely there was something I could do, some errand or diversion to occupy me. But the sense of immobility crushed me. Even the next moment was contingent, imperiled. One building rose above the others on the West Bank campus—the social sciences tower, at the far end of the mall. All the dark square windows. Someone had witnessed, from the social sciences tower, my father's last, imperfect crossing. Surely no one was there, on this holiday. But—someone is always watching. But—what does it matter? Still— My brain told me to take cover, but refuge was distant.

On the other side of the river the avenue sloped upward, and a footbridge crossed over it from the main campus to the student union and the hospitals. A figure appeared on the footbridge. This was quite a distance off, but I had given Sylvia a month of intimate attention and I knew—I would know from any distance, any angle—that this was she. Without thinking I started toward her, began jogging across the long bridge. I waved and yelled—no use. The wind was too strong—she would never hear me shouting. I was halfway across the bridge. She came off the footbridge and made her way by a series of sidewalks and stairs toward the research labs' entrance, and she moved in and out of my vision. She approached a set of glass doors. She reached in her bag, brought out keys, considered several and then fitted one to the lock and pulled open the door. Her actions were so calm and deliberate, it was as if she were teasing me. I was near the end of the bridge but still some distance from her. The dark glass door closed on her, she vanished, as if beneath the surface of dark water.

I slowed to a walk, out of breath, lightheaded. The hospitals complex, constructed bit by bit over decades, began below to my right and stairstepped up the riverbank in a mish-mash of crags and nodes till it rose to modern twin towers some distance ahead of me. Behind the baroque façade, or beneath, the lab complex extended well down and into the steep bank. I had been inside once or twice; I remembered it as a massive, labyrinthine place where you could easily lose your way in the maze of identical antiseptic hallways and stairways. I didn't even know where Sylvia's lab was. All of which appeared now quite academic, since I couldn't get in. I stopped at a little distance from the doors, and regarded them tentatively, as if they posed some hazard. The dash after Sylvia had given me momentary purpose, and now left me here stranded on the far bank, with the river between me and the car. With nothing else to do, I sat down on a concrete ledge near the door and lit a cigarette.

I looked west now. The bridge, the West Bank campus, the social sciences tower. I regarded the scene impassively. The bridge, it seemed, should be symbolic, somehow. But if it was a symbol, of passage from one thing to another, I didn't understand it: not the meaning of the passage, not the one thing, not the other. The bridge was the last place my father had stood on earth. It wasn't on earth. It was an artificial construction, up in the air, over the river, between one thing and another. And always I returned to it without understanding. So would I go round and round my whole life, and never move beyond this? But it wasn't neat and circular, no consoling pattern could be seen. Rather it was like a scrawling looping doodle where lines cross and recross and veer off wildly and backtrack till both shape and direction blur in a confused smear of ink, so the mental record of my lifetime was like a page, or a whole book, filled with scribbles both ugly and indecipherable.

Now my scribbling brain wandered back across the bridge

(the bridge was empty, as my father had left it) and to the social sciences tower, and up several floors, and into a corner room with a dark oak conference table surrounded by comfy chairs, and wooden bookshelves around the two interior walls. An unexpectedly cozy room in the modern concrete building. A nice place to watch the blizzard that is about to batter the city (the brain has spanned years as well as water, and journeyed into winter; the brain is scribbling back over scribbles of seven and a half years ago); a good place to watch the coming storm, to see the hard fine snow blowing across the mall and down the bridge, and across the gray slick surface of the river ice. My brain sidles up to a man standing at the window—we don't know if he's young or old, a student, professor, custodian, office worker—and puts its insubstantial mental arm across his shoulders, and looks down with him. The bridge is empty, it always will be. The bridge is a vacuum that my father made falling. It sucks all thought and feeling away into a cold, still place, and never will release them. The man contemplates the vacuum, the void, or his eyes are directed that way, but they focus on nothing. His arms hang at his sides, the hands slightly curled, for he has just reached out reflexively and grasped at nothing, and he feels the emptiness they hold like a film of oil that will not wash away. *Hey,* says the scribbling brain, gently. *What did you see? It's okay. Tell me what you saw. Slowly, one step at a time. Tell me only what you saw, don't embellish, don't get excited. It wasn't your fault. There's no one there. It's no one's fault.* The brain grasps the shoulder, comforting, gives the slightest urging shake, to give him confidence. *A tall man, bearded, in a blue down parka,* the brain begins, to get him started. *No hat, no gloves, though it's incredibly cold, it's been cold like this for a couple of weeks, the usual January plunge, but longer and deeper this time. Maybe you didn't notice that he wore no hat or gloves. When did you first see him? Was it just a streak of blue falling toward the ice, so you didn't even believe at first that you'd seen it?* But since the bridge is empty, so crushingly empty, the man can find nothing to hold on to, nowhere to begin. He stares on and on without answering or even turning his head. *Okay, it's all right, never mind,* says the brain, a little condescendingly, as if to a child, and lets go the man's shoulder, and drifts away. Maybe this was the wrong man, anyway, maybe he wasn't one who would stare at nothing or feel the emptiness in his palms and fingers. Maybe the man in the window said, *Jump, jump, do it!,* though no one could hear. Maybe his own brain reached out and shoved my father off the bridge. Impossible to tell. Didn't matter. But the vacuum drew everything toward it, and when it closed sent out a rippling backwash, and the man in the window felt its waves. We don't know if he relived this witnessing again and again in his dreams, and never spoke of it aloud, or if he reported it in loud tones whenever conversation lulled, since it was surely about the most amazing thing he'd ever seen, a full professor, a poet of small renown, dropping off a bridge. Don't see that every day.

While my brain doodled my eyes watched the doors, and on their dark surface two dim figures now appeared, gaining shape as they neared, just as if they were rising from murky water. I hopped off the ledge and walked toward the door and reached out for the handle as they came to the threshold and one pushed against the bar, swinging the door out toward me. I took the edge of the door and held it open for them, smiling, nodding as they nodded back and passed without comment. I went inside.

I had spent considerable time loitering around the University in the last month, looking for Sylvia here at the hospitals complex or at her house across campus. We both had schedules that made planning difficult, and neither of us was easy to reach. Sylvia was constantly in and out of the lab, as I was at the newspaper office. She had evening classes; I had dull civic meetings to attend and report. The boarders at Sylvia's house shared a single telephone on the main floor, and it regularly went unanswered. Sylvia lived in blissful—I guess—isolation, high above the chiming telephone or doorbell. Thus our meetings were often haphazard. We had no plans for today; we never had plans. This way, if ever we met, it seemed like fate, all over again. New lovers are atavistic in the extreme, looking anywhere for signs and omens to confirm their hearts' choice, while the heart still teeters in uncertainty. Destiny, give a nod, won't you? To give destiny a leg-up I had wasted a lot of time waiting on the chance of finding Sylvia here at the lab or at her house; and it was not comfortable, haunting this place where I had once belonged but where I now felt old and awkward and unconnected. I suffered the discomfort because I had come to live for the hours I spent with Sylvia in Sylvia's little room, which had lost none of its magic for me since the first time I was brought there, a month ago. What were a few hours, what was a little unease? At that time I think I would have stayed chained to the fire escape railing, just to be there when she returned.

If I met Sylvia after work it was always easiest to meet her at the door. Now that I was inside the building I didn't know which way to go. Corridors led to the right and left from the smallish entry hall. Ahead was an elevator and a door into a stairway. I went into the stairway of harsh incandescent lighting, off-white walls, tubular steel railings. I went down almost on tiptoe, fearful of being caught and expelled. One floor down I stepped out into the hallway. It was carpeted, lined with glassed offices all dark. I went back into the stairway and descended two more floors. It looked more scientific here. The hall was an off-white monotony under the flickering pallor of fluorescent lights, with metal doors of the same color at regular intervals. The doors were numbered, and I was in the 400s, about in the center of the complex, I guessed. That seemed as good a place as any to start. Inevitably, I suppose, I thought of a rat dropped in the center of a maze. Sylvia was my reward, my cheese, and I had to somehow scent her out. Was the incentive great enough to stir my atrophied instincts? I tend to disapprove of behavioristic models. Nevertheless, when in Rome. . . .

The identical doors were here and there distinguished by small personal touches—an erasable memo board, like you'd find on a dorm room door; cartoons featuring scientist gags, so many and so repetitive that they struck me as a self-defeating effort to prove that scientists can take a joke; a child's drawing stuck up by some proud parent. Some of the doors carried radioactivity warnings—the pie of the three truncated black wedges floating on a sickly greenish-yellow field. Sylvia's lab would be one of these, but they were many. I was too timid to knock, and that reduced my odds to the chance that Sylvia left the door open when she was working. Was this place as deserted as it seemed, or were there people behind these doors, working feverishly at their strange arts?

The first floor was relatively simple, three rectangular blocks that I covered in a few minutes. The next level down was more difficult. It seemed that two buildings or two stages of construction met here. The corridors went every which way. They jogged and dead-ended and split in T's, and while I concentrated on tracing my way through this maze, how could I be sure there wasn't a hidden cul-de-sac I had missed? The rat in the funhouse analogy which had occurred to me cursorily above now became more cogent. Which is to say—what? I was sure I'd seen the cartoon on this door before, but wasn't that on the upper floor? This door was gray where that was beige. How had I managed to descend from the fourth to the sixth level? I remembered the view of the jumbled exterior and its intimation of impenetrable depths. The complex hummed here; I had noticed the low whirr of the ventilation system on the floor above, but there it had seemed soothing. Here it almost throbbed, and it was something I felt more than heard. And there was an odd smell here, an antiseptic smell mingled with animal funk, like a veterinarian's office. I thought I might be imagining it, till a volley of barking sent me bolting across the corridor and into the opposite wall. The barking subsided to whimpers, yips, and growls as I leaned against the wall to catch my breath. I gave a little panicky laugh, and continued down the corridor. The outburst had set off a chain reaction and as I continued along, ever more anxious and claustrophobic, I heard an unnerving symphony of feline yowls, monkeys' shriek and chatter, more barking, and even, I thought, a cow's deep plaint. It was a jungle down here. I followed corridors and turned corners indiscriminately, and my plan of systematically covering each floor went all to hell. When I stopped to orient myself I found that I had no idea where I was or from whence I had come. I had dreaded being discovered; now I longed for someone, anyone, to find me, arrest me, throw me in a cage—just take me out of this bedlam. I found another stairway. It only went down, and I followed it. Finding Sylvia was the imperative that cut through my panic. She was in here somewhere. But then it occurred to me that she might have just come in to check something, done it, and left. I tried not to dwell on the thought.

By the time I had made my way through the next level, which was composed now of three different sections, and at least two numbering systems—so I was at the second and fifth levels at the same time—with no sign of Sylvia and no confidence that I would ever find her, the imperative had become simply getting out. I remembered the elevator I'd seen when

I came in, but I hadn't seen any elevator since, which struck me as sinister. I decided to be logical and see where that would get me. I considered what I knew about the lab complex and its relation to the outside world. It was connected to the big university hospital, though I did not know where, or how. It was connected to the main campus by at least one tunnel that ran under Washington Avenue—when I worked on the University paper we had printed a map of the elaborate system of tunnels that led throughout the campus, and yes!, I had helped research the story, as almost everyone had, and that was one of the times I had been in the labs, I had walked through that tunnel. And it all came back to me, the stuffy, overheated air, the echoing footsteps, the bang and hiss of steam pipes, the eerie rumbling vibration as traffic passed overhead. My sensory recall was acute, but as to where the tunnel entered the labs, I hadn't a clue. The labs were also connected to a large parking ramp, at the lower corner away from the bridge and near the river. That had to be it, my way out. The ramp didn't reach as high as the lab complex did, but I was sure I'd descended far enough that the ramp and the complex now met. I oriented myself as best I could and moved in the direction where I thought the ramp should be. A door with one small window showed gray daylight. I pushed open the door and walked out into the brisk cool breeze. I found a stairway and went down to the entrance, walking through the nearly empty ramp left open without an attendant on the holiday, and out to the bright sun and onto the river road. With the exception of certain last-days-of-school, I don't think I had ever felt so free.

It was a steep, long walk up the river road to the bridge, but a lightness in my legs, in my entire body, carried me effortlessly up. The day had changed: the blank dead light had become fresh and comforting; the empty wind was lively and brisk, sparking the small new leaves to bursts of color and sound. I forgot, for a while, that I had failed abysmally in my search for Sylvia.

I turned right onto a short drive that led sharply up to Washington Avenue, then up a set of stairs to the pedestrian level of the bridge. Back again at a point of decision, my good mood dimmed. I looked back at the lab complex tumbling down the bank; I pictured its devious catacombs. I looked toward the door where I had entered, and imagined a small figure appearing on the dark glass, gaining shape like a photograph developing. The illusion was powerful; and then it was no illusion: the door burst open, out stepped Sylvia, alive, three-dimensional. She didn't notice me at first, and for a moment I was unable to speak, and it seemed that the proper closure to this episode was to let her walk away without seeing me, and then I would turn to return across the bridge, as if only in that way could the destiny of this day be consummated. Then I found my voice and called out—

—Sylvia!—

—with a note of desperation and longing entirely unintended. She turned already smiling—had the plaint in my voice survived through the wind? She stood still while I walked over to her. We hugged, and only then did she seem truly real to me, her live small body so certain in my arms.

We sat down on the ledge where I had waited earlier. We shared a cigarette. I laughed as I told her of my futile search for her, the wailing animals, my eventual escape. She tried to describe where her lab was, but I couldn't put it in context. She'd had the door open, but somehow I had missed her.

—I still get confused down there sometimes, she said. It's easy to forget what level you're on. I get around automatically now. If I have to stop to think where I am, I'm in trouble.

As we sat the magic mirror door produced another figure.

A man stepped out into our dimension. Sylvia looked at him and said:

—That didn't take long.

—The counter broke down again, the man said.

He looked to be in his early thirties, smallish but sturdily built, light brown hair receding, a thin mustache, round black-rimmed glasses on a slightly babyish face. He wore faded jeans and a gray oxford-cloth shirt and running shoes. I stood to shake his hand as Sylvia introduced us: Robert, her boss at the lab.

—I'm nobody's boss, he said. Just another lackey. There's a boss above me and one above him, and who the big boss is god only knows. The man who gets the grants. We only know of him by innuendo.

He was pleasant but a little distant and theatrical, not actually friendly. He mentioned something about summer schedules, as the term would be over in a couple of weeks, then he left, saying he was going to go for a run—the way he said it made it sound like a moral judgment as I dropped my cigarette and stepped on it.

Sylvia didn't ask how or why I'd come here, and her acceptance removed the air of strangeness and menace from my dash across the bridge, my witless wandering in the labs. Now it was just a pleasant sunny day, the last of May.

—I just talked to Carla on the phone, Sylvia said. We were going to do something today. I said I'd come to her apartment.

I told her my car was across the bridge. She didn't need to go back to her room. We began walking together across the bridge.

When we reached the midway point of the bridge I stopped.

—It was just about here, I said.

—Your father.

—At the end of January.

—So cold that year.

We looked down at the river that moved and shimmered under wind and gentle sun. The dark water showed no depths, and was not a magic door—no man or phantom rose to its moving surface. But, it was a different river now, not ice-locked, gray, and the wind was kind and soothing. I willed myself to not imagine winter. Sylvia took my hand, and we turned and walked on.

Best Angel

We picked up Carla and left the city on the freeway headed south, the same road I had driven earlier that morning. But such a different road. I remembered that dreamy drive, and I smiled as if I shared a precious secret with someone.

We floated down a long hill toward an iron trestle bridge that curved over the wide marshy bottoms of the river of many lakes. The long hill upward, ahead, promised free and open land at its crest, even though the suburbs now stretched well beyond the river and for some miles yet we would pass subdivisions and office parks rising up from the cornfields.

We sat together in the front seat, Sylvia in the middle. The AM radio tinned and buzzed out chipper pop songs. It all sounded good—the latest insipid rendering by the latest teen sensation, or some plasticized soap opera star, was a psalm of pure joy and delight. By the end of the day we would know *all the words* to a dozen new songs by heart.

The cavernous Bel Air felt empty even with the three of us in it. And though it was large, it was relatively light, a great tin can on wheels, and the V-8 fairly threw it over the hill. We hurtled past struggling subcompacts with a decadent, oblivious sense of American superiority. We were burning a gallon of gas every twelve miles.

The view at the crest was worse than expected. The hills had been cropped and flattened by miles of identical townhouses and duplexes, each distinguished only by slight variations in the sickly hues of the vinyl siding. Spindly saplings on the bleak expanses of newly laid lawns spoke a hope in the future that I found unimaginable. The tiny trees were tethered all around with lines staked in the grass, as if to steady them in their stupendous growth. A shopping mall sprawled off to the right, its great desert of parking lot nearly empty on this holiday. Ready, set, go, it all said: Bring on the future. Round two—the older suburbs lay staid and dingy behind us.

Sylvia twisted around in the seat and pointed out the rear window.

—I used to ride my pony up there, she said. I rode in those hills. The Feltls must still have the ranch.

In the rearview mirror I saw a tall round grassy hill. "Ranch" sounded a bit grand.

—My sister rode there too, Carla said. But I was never too horsey.

—And over here was all just empty, just farms, Sylvia said.

The suburb where Carla and Sylvia had grown up backed onto the river at the long-neglected bluffs which were now prime real estate, "executive homesites." It had seemed so distant back then, the frontier.

—I can't believe I've fallen in with suburbanites, I said.

—Survivor of the hellish inner city, child of the ghetto, Carla said. Where you grew up, they'd call it "an enclave of the very rich," if they had to call it something, in *Time* magazine, say, or on TV.

—It is the inner city, I said. You could see downtown just over the trees.

—I can *see* downtown from here, Carla said.

Which was true—the freeway gave a straight clear view to the tall buildings, now miles behind. But we were on high ground.

We exited the freeway onto a county road that ran straight through the rolling farmland, tracing a pleasant line over gentle hills and dips which the freeway disdained and flattened. The stars and stripes flew above tidy farmsteads. Sylvia had said to turn off here. She had a destination in mind, perhaps; at any rate she would be our guide.

We made our way southward over blacktop roads that neatly gridded the countryside. So straight were the roads, any deviation seemed a virtuoso turn of the roadmaker's imagination. How inspired, after so many straight miles, to give us this curving lakeside stretch to consider! Approaching a small river gully, the road appears to carry on straight across, but as we crest the ridge—bravo!—he sends us downstream a quarter-mile for crossing, and back up the other side the same distance, to return us to the expected way. Author!

We had been driving about an hour when the road presented ahead a steep grassy ridge, and with perfect understatement took itself away to the left (quite sharply, but not *too* sharply), leaving us to contemplate the rarity of elevation in this low landscape. The ridge rose in three steps, each finishing in a rounded top, with the first two dipping to rise again and the ultimate curving away into the hard blue sky. A white wooden church was sited atop the first rise, and a graveyard climbed the second one behind the church. I pulled into a gravel parking lot at the base of the hill, and we got out of the car.

From a distance the church had looked freshly painted, a bright clean white in the strong sun. The entire scene had spoken of peace and order: the fine symmetry of the church with its short sharp steeple and its neat row of modest arched windows down the clapboard side which faced us; the building's noble prospect, seated midway between heaven and earth; the winding stone walk which led up to it from the road; even the cemetery, ranged behind on the grassed rise, had looked quaint and cheery.

We'd been deceived by distance and light. Crabgrass inhabited every joint and crack on the stone path, and the stones had lost foundation and gave treacherous footing. Without the sun to spread what little white remained on its boards, the church was gray and worn, the exposed clapboards succumbing to sun above and wet below. One window was covered with a black board, the rest with grayed plastic. The double front doors were secured with a padlock and chain threaded through the door handles, but the handles tilted out from rotten screw holes. We walked around the church in silence, like some sort of inspection party, or a new landing of colonists arrived to find our countrymen mysteriously vanished. One large elm near the front of the building gave down a soft green cast from its bright new leaves, as if to mock our blasted hope.

Never mind. We hadn't come to worship. And when the sun emerged from behind a passing cloud the scene regained the rickety charm which decaying country things seem to hold for city people.

—Watch your step, I said as one of the pathstones gave way under my foot.

You'd have thought I'd have noticed, at any distance, how overgrown the church cemetery was, and that cemeteries aren't supposed to look that way. The military cemetery had been a grimly impressive spread, its orderly rows of gravestones, so white in the sun against the green spring grass cropped and tended like a fairway, stretching on for miles, it seemed, disappearing in a wave over the rounded edge of the bluff. Grand oaks and dark trimmed cedars gave relief from its mortal monotony. As I drove past, the lines of stones jerked my eye this way and that to follow the shifting, switching straight, diagonal, straight—a dizzying domino geometry of the dead.

The graveyard here began just behind the church, and rose and spread up and over the second rise in an unregimented sprawl, with stones splayed and kiltered on the uneven hillside.

—Ouch. Shit, Carla said.

I turned to see her bending, rubbing her shin. She tramped down the grass with her foot to show a section of low spiked iron fence, and once she had pointed it out I could see the small square yard it described on the hillside. We all looked, and it seemed to strike all of us at once, how far the graves had overspread their hopeful bounds, the quaint conceit of the tiny fenced yard which contained perhaps one tenth of the plots—contained not very well. We laughed. Sylvia put on a voice of cleric authority and said:

—We'll have just this many dead, and then no more. Once the graveyard's full there'll be no more dying.

—I guess you can't imagine how many people are going to

die, Carla said. But you'd think religious people, at least, would have come to terms with that.

—Maybe they were some kind of apocalyptic sect, had a date for judgment day, I said.

—Or maybe they just ran out of fence, Sylvia suggested.

We wandered along the hillside, looking at the headstones secreted down in the grass, trying to read the inscriptions on the red and gray granite stones that had long since lost their sheen and were weathering to dust, as if the stones had been designed as hourglasses of memory. The family names were Norwegian, German, a few Czechs; read one after another they sounded stolid and severe, it was a no-nonsense graveyard. But how young these people had died, how common was death in those days—not that long ago. Just being born or giving birth posed mortal risk—the stones attested to that. And then there were smallpox, influenza, tuberculosis, the croup, the grippe, and what-all; a potpourri of diseases whose names were now old-fashioned and quaint. But sure, they'd been deadly. You could cut yourself shaving and die quite horribly of tetanus. I came across a family plot of red grainy stones—the Sturdevants—where around the central monolith all the circling stones had fallen outward like a daisy's petals. A mother and four children had perished in an eight-year span, the mother and the last child together, followed shortly by the father. The hardy grandparents had lived into their fifties; brothers, sisters, and in-laws kept death a constant factor. The Sturdevants had flourished (if you could use that verb for a family of such stupendous mortality) in the late 1800s. What of the latter-day Sturdevants, and the descendants of all those buried here? Was it too much to keep up remembrance as the shadow of the dead grew longer and darker over the light of the living? Was it some kind of pathology to say to hell with these old bones? It wasn't something I could judge. We are so protected, we think nothing bad should ever happen. I counted funerals and found that I had attended three—two grandparents dead of plain old age, one father by his own force.

A bridge again. It was, after all, Memorial Day. And here I was at my second boneyard of the day, neither visit by choice or design. And so pondering this coincidence (ah, coincidence), I climbed halfway up the third tier of the rise and sat down in the tall grass.

Carla walked below me along the spine of the second rise. I could ask her the meaning of these latest coincidences. Below her Sylvia wandered among the headstones. She looked carefree and light, with her hands in her pockets, bending down now and then to brush aside the grass to read an inscription. And she was singing, sporadically; I just caught bits of a delicate tune, "Annie Laurie," perhaps. Carla was watching her too, and I watched Carla. Then Carla felt my gaze on her, looked up at me. She looked away, turned back to Sylvia. But my eyes had caught hers when she looked, she had to look back. She teetered on the ridge between me and Sylvia, then she fell toward the higher ground. She waded through the grass to where I sat.

—What ho, Prince, Carla said. Contemplating mortality in this evocative setting?

I waited a moment before speaking. I don't like to leap on allusions.

—I'm not Hamlet, I said. I'm . . .

I'm—what's his name? I almost said: I'm your funny Valentine.

—You're who?

—Nobody, just a guy. We can't have that Shakespearean grandiosity. Tragedy is passé.

—Oh, tragedy is always popular, isn't it?

—Well, I guess, I said, and we let that topic roll away into the grass where it was lost like a golf ball in the long rough.

Carla pulled out a stalk of grass and dangled it from her teeth. Meantime I had lit a cigarette. We both were watching Sylvia, who was paying us no mind.

—I don't think I really understand about the car, Carla said.

Since Carla had brought Sylvia and me together she had receded in my life, somewhat like the catalyst in a chemical reaction. But she had been spending time with Sylvia, so the analogy is perhaps imprecise.

—I don't quite understand it either, I said. I bought it for a dollar, which I haven't paid, or I borrowed it but I mustn't bring it back, or I got it in trade for cleaning up her grandmother's yard, which is a mess, which I haven't cleaned up. It was the first time I saw her after the party.

—That's a pretty eventful first date, Carla said. I guess that sealed your fate.

—What do you mean *fate?* What does *fate* have to do with it?

—I just mean, you go on a first date, your date's grandma gives you a car. Where do you go from there?

Where indeed. I had trouble connecting Sylvia with "your date," though.

—Do you say that coincidence and fate are the same thing? That coincidence is part of fate, fate's medium, or its manifestation? What?

Carla laughed, threw up her hands.

—How do *I* know what *fate* is? she said.

—I thought you had a theory.

—Well, yeah, a *theory.*

—But what isn't coincidence? I mean, except for the few

things in your life that you actually plan, or intend to do, like doing your laundry or getting your hair cut, what isn't coincidence? If you plan to get your hair cut, then you go and get your hair cut, that's not a coincidence, but if I happen to find Sylvia at the University, and take a drive, and wind up at another cemetery on Memorial Day, after the bridge again, the bridge my father jumped off, then that's coincidence, and its *meaningful?* Then this conversation is full of omens, since it began with the original coincidence, and so are you, Carla. What do you *mean,* Carla? And the whole rest of your life, from the moment of the first coincidence, is part of it, it's all nesting boxes, coincidence to the *n*th power. And what if you missed the first one, and don't know where it started, like coming in late on a confusing movie. You never catch up.

Now I had Carla's full, if befuddled attention. She gaped at me.

—I've lost you Bryce. I think you're going at the whole thing much too rationally. My idea is that you should leave yourself open to coincidence, let events lead you, sometimes. I think a lot of people don't. They get things a certain way, a job, get married, buy a house, say, then they just go along with tunnel vision, refuse to consider all these possibilities that might appear if they would just look. But it's not like everything is omens, it's not like, cut open some animal and read the future in its guts.

—But it might as well be, right? Because you find coincidence where you want to, you follow possibilities to which you're already inclined. You confirm your prejudices.

I was thinking about Frank, who revealed to me what I already knew, and about the secret desire. I didn't know if the secret desire was a positive force, or if maybe it was just another face of the antagonist of bliss.

—I suppose that's true. But at least if you see the possibilities, then you can choose, Carla said. It's not animal guts, and it's not a system, like math.

—Math isn't a system like math, I said. In math there are imaginary numbers, things that exist but can't. In the calculus of imaginaries. Makes sense, doesn't.

—Well then maybe it is a system like math.

—What is?

—Math is. Life is. Maybe Sylvia could enlighten us. She understands math, at least.

—I wouldn't ask her, I said.

—Why not?

—She's the heart of all the coincidences, isn't she? She'd just drag us deeper into the muck.

—I forgot: you're in love; you're obsessed.

—Did I tell you that?

—Who needs telling?

—Wasn't it "smitten, enthralled"?

—It progresses by degrees.

Then we stopped talking to watch Sylvia, who still strolled through the derelict memorials as if through some dreadful garden; but to her it was apparently not dreadful. Now she was singing "Wild Mountain Thyme." Then of a sudden she turned to face us, and flashed an exaggerated smile, and waved. We waved back, and then Sylvia turned away again to follow her own amusement. Carla said,

—Whose idea was it that you take the car?

—It was her grandmother's, mostly. Entirely. I didn't have a choice.

—Oh, you had a choice.

—What difference does it make?

—I just wondered what Sylvia thought about it.

—She was opposed. She thought Gran should keep the car. Why?

—Nothing.

—What *is* your point?

—No point.

—You mean you think Sylvia didn't want me to have the car?

—How should I know?

Sylvia shouted from down below:

—Hey! Behold!

She had found a large stone angel that had lost its head, and climbed up behind to put her head on the blunt empty neck.

—I bring you tidings, or something, Sylvia yelled.

Carla and I walked down to where Sylvia posed. Carla had brought her camera, and she took a picture of composite angelic Sylvia. Then Carla and I took turns having our pictures taken on the headless angel, and we staged a "Best Angelic Expression" contest. I didn't know what kind of angel to be, wrathful or beneficent, and the resulting look of witless confusion got me hooted off the pedestal. Sylvia easily won the contest, maybe because she had spent so much of her childhood studying that book of favorite saints. By the end we were sprawled in the grass, hysterical, sick with laughter. I was wishing we hadn't stopped. I was wishing we had found here a reverently tended church graveyard which radiated a calm acceptance of mortality and promised posterity's tender remembrance. Then we wouldn't have wound up unhinged in the grass, laughing like idiots. And I kept wondering why Sylvia didn't want me to have her grandmother's car.

A shout cut through our laughter, startled me, though at first I thought I imagined it. It came again:

—Hey, you . . . *something, something.*

We sat up in the grass like startled animals.

—Hey, you kids. Whadda you think you're doing up there?

We rose to our knees and looked down the slope like prairie dogs surveying a dangerous plain. A man stood near the church. He was waving his arms in an impatient beckoning gesture.

—What the hell do you kids think you're doing? he shouted.

The question, I took it, was rhetorical. And since I didn't exactly think of myself as a kid I glanced quickly around, half-expecting to find a gang of teenaged vandals menacing the graveyard. At the same time, I felt the plummeting pang of the child caught at a boisterous but harmless game: "*Geez,* I wasn't *doing* anything." The kids he was yelling at was us. Was there any law against frolicking in cemeteries? Could we just duck back down in the grass and ignore him? The man had the air of someone who would not take well to being ignored, there was a sense of self-imposed authority about him. We all three looked at each other, shrugged.

—Uh-oh, we're in trouble now, Carla said.

She let out the last of her stifled laughter, which surely did not endear her to the man down the hill. We had to make our way carefully down the hillside, to avoid tripping on hidden stones, and the man stood shifting impatiently. He even looked at his watch once, as if we were taking hours to come down. In glances up from the treacherous ground I observed a man near fifty, with slicked-back thinning hair, a harsh set to his small eyes, wearing a yellow knit sport shirt and gray dress pants with wide cuffs that drooped over his brown dress shoes. He looked like a mean guy, at the height of his meanness, when he'd seen enough years to confirm the correctness of his outlook, but not enough to mellow him. I thought: I'm glad he's not *my* father. He was clearly somebody's. His car, parked behind the Bel Air, was a fairly new American two-door, kind of sporty, really, white with a brown vinyl roof. I realized that my car didn't do us any credit in his eyes. From its original status as the apotheosis of respectability it had become the symbol of bombing-around-looking-for-trouble kids. But I think we gained a small advantage as we came to focus in his eyes and didn't fit what he expected. I tried to capitalize on his confusion with a wave and a hearty "Hello!" as we neared. But that backfired—only the guilty would try such a bald, pathetic ploy.

As we were stepping over the low fence he said:

—This is private property. What do you think you're doing?

Any answer was the wrong one. Carla said:

—We just stopped to have a look and take a walk. We really weren't doing any harm.

Sylvia tried a different approach.

—Whose property is it? she said.

—It's the church's.

—Did you go to church here?

—I'm a *Methodist,* he said, as if offended. —Kids come along here all the time and knock over the headstones.

Then we were no longer grouped with the destructive kids, and a truce was in the offing. But he got in one last lick:

—It's Memorial Day. You could show a little respect.

We agreed, we sort of apologized. I think the man was a little disappointed that we weren't a bunch of young country vandals whom he could flail hell out of. This was a man who knew what was right and what was wrong and what was what, but he didn't know what to make of us. He looked us up and down, and then fixed for some time on Carla's chest. I was thinking I ought to rise to defend her honor, then I realized he was puzzling over her T-shirt. Under an open pink cardigan, Carla wore a faded black T-shirt emblazoned with a large ornate red "A"—a promotion for a public TV adaption of *The Scarlet Letter.* The man appeared to understand that there was some reference here, but the light wouldn't pop.

We'd been properly chastened, professed guilt in spirit if not in deed. He in turn had lost the fire of outrage which had made him loom. So we lingered in an awkward standoff of people strange to one another. He could have left, that would have been best, but he seemed reluctant to go. We were in no hurry to get anywhere, and as we were three it was more awkward for us to coordinate a graceful retreat. After some moments of shuffling silence. Sylvia said:

—Do you farm around here?

The man put his hands on his hips, and pressed his lips together as he shook his head.

—Oh, no no no. Nope. My folks did. I didn't want to take it on, neither did my brother.

He turned and pointed across the road where an empty pasture surrounded an abandoned farmhouse.

—My brother-in-law had that place, ran a dairy. He went belly-up and moved to Arizona for a job. Now he's got a place with a swimming pool, but his yard is made of rocks. His whole neighborhood is nothing but snowbirds, he says. We'll probably visit him next winter.

I had a flash of this pale man in baggy bermuda shorts by his brother-in-law's swimming pool in an Arizona subdivision, his hairless shins and basted scalp going pink in the desert sun. He sits uneasily in a folding lawn chair, with a can of beer in a foam holder to keep it cold. He tries to believe this is paradise, but he can't get over the fact that the lawns down here are gravel or Astroturf. A teenaged daughter, with braces on her teeth and a "Sunny Arizona" T-shirt, is burning the backs of her legs on a towel poolside. My heart warmed. Then I imagined whole herds of pale northerners—the "snowbirds"—

in polyester and nylon, wandering lost and dazed among towering saguaros and cattle skulls and rattlers. My heart sank.

Carla asked the man what he did then, if not farm, and he said he worked for a well-drilling company run by his cousin.

—Over in Lindquist, he said, throwing this thumb out toward the southeast.

—My uncle has a farm near Lindquist, Sylvia said.

She mentioned the family name, which he knew.

—Stenmark, sure sure sure. We did some work for him a while back.

So now we were practically like family, and we could all be friends. We asked about the church, and he said it had been absorbed by a bigger church in town. The cemetery had been closed for many years, and it was too difficult to maintain, with the stones going every which way and falling over on the steep hillside. A pretty spot for a church, but no kind of place to be dead. Nowadays they set the stones flush in the ground, for easier tending. There was talk, from time to time, of moving the—he searched for the respectful term—*remains* to the new cemetery. He gave a little laugh. He seemed embarrassed to have broached such a morbid topic.

—Some of them, you know, some of them were *scalped,* he said, his voice suddenly excited, and he directed our attention to the sprawl of markers, as if we might observe the bony bald *remains* sitting up on the hillside. —So they say. No more Injuns around here now. Long gone.

He guessed we were from the city, and what were we doing, just out for a joyride? He'd owned a Bel Air like mine once, or maybe it was a Biscayne or Impala—same color, 1962 blue. And what was the city like, now, pretty nice? He and his family usually came up for the State Fair. We said it was okay, it was changing, and he nodded his head knowingly.

—My brother's girl went up there to the U. Studied psychology or some damn thing. She bought a little house on the south side, but now the neighborhood's going black.

He turned toward the north, and we looked with him, and we studied the horizon, as if the encroaching threat, all these people who were *not like us,* might appear there like those long-gone Indians on a Hollywood ridge.

In a moment we all four started down the path toward the cars. The man assumed a supervisory air again, and his good-bye as we climbed into the Bel Air was curt. The position of our cars, his behind mine, and the initial tone of our encounter, made it feel too much as if we'd been pulled over by the police. And I didn't want him following us, so I waited to let him pull out first. He was waiting, too. He adjusted his visor and mirror, checked his seat belt. Our eyes met in my rearview mirror. He started his car, revved it sharply, took a look over his shoulder, and pulled out onto the road with a harsh spurt of gravel. He accelerated quickly down the empty road, not looking at us as he passed.

I started the Bel Air which didn't need a key, and eased decorously onto the blacktop.

Fishing

Sylvia and Carla laughed away the tension left from the encounter with the well-driller, describing what had happened and what they had thought.

—I honest to god thought it was one of my uncles, yelling at me to keep out of something at Gran's, Sylvia said. A horrible flashback. That awful tone of voice. It makes you feel like a dog.

—I know it, Carla said. How could he pull those strings so easily? We're all guilty kids deep inside.

—Very profound, Carla, Sylvia said. Does it all go back to potty training?

—Shut up, Carla said, and pushed Sylvia, who bumped my shoulder, and we swerved, and they laughed.

—Sorry, Sylvia said, leaning briefly toward me and then turning back to Carla.

—What a sleaze, Carla said. Did you see the way he kept staring at my tits? Not that there's much to look at.

—I think he liked you. You're exotic, not like us run-of-the-mill Scandinavians. He's probably telling lies about you right now. He's probably dreaming of getting you out behind the polka palace.

—Most of the people around here are German, I said.

—I *love* to polka, Carla said. Maybe I'll call him up.

—In the whole state, for that matter. This Scandinavian shit is all PR.

—You'd be the queen of the dance. They'd be fighting over you for the schottische.

—That truly is my fondest dream, Carla said.

They laughed, their laughter going silent as it grew. Sylvia took Carla's hand, they leaned their heads together and shook with silent laughter. I said:

—He was looking at your T-shirt, your adulterous "A."

I don't know why it came out that way. They stopped laughing and pulled apart. I braked hard for a stop sign.

—Do I turn here? I said, too sweetly.

—No. Go straight, Sylvia said in her high voice, quietly.

The farm was deserted. Where the hell did everyone go on Memorial Day? And why, at any rate, were we seeking out Sylvia's terrible shouting uncles? I had a lot to learn about families, I guessed. But that was the other side of the family, I remembered. Sylvia's family came from farms all over. These uncles, on her father's side, were gentle on top, with their Nordic rage and madness tucked tidily away inside. From time to time some farmer, or farmer's son, or even farm wife, would murder the entire family with an axe or a shotgun. That was one risk of the circumscribed emotional life.

The farmhouse was set among lovely oaks. The fields, stitched with infant green, swelled softly to a line of distant trees—the property's bounds, and a small fast river ran there, Sylvia said. There was a picnic table in the shady yard. Birds called, a fat orange tabby drowsed on the front steps and paid us no mind. The front door was open, inviting breezes. A scene of peace and trust. Then why did it make me think of axe murders? Somewhere, in one of these sheds scattered about the place, there was a lot of dangerous hardware.

We had picked up some sandwiches, a bag of chips, a six-pack of three-two beer at a convenience store. We eschewed the civil picnic table and set off instead for the river, walking in a section of fallow ground at the edge of the sprouting fields. Sylvia led and Carla followed and I brought up the rear. Carla did indeed look exotic, especially here in this scene of rural Americana, amid the dirt and growing things. With her shocking pink sweater she wore voluminous pink-and-black striped trousers and black cotton Chinese slippers. Her features were subtly Semitic, her eyes pale blue, her skin light olive. Her fine twisting hair, light brown showing reddish gold in the sun, lifted off her shoulders in the breeze. I pictured her lounging among silk pillows, not tramping through a cornfield. She was enjoying herself, though it was clear she had not walked in cornfields much—her balance was always questionable, and she held her arms slightly extended, as if walking a tightrope. Sylvia, in jeans, a red cotton sweater, and tennis shoes, negotiated the uncertain ground with surety, and with just the right amount of agrarian awkwardness, humility, as if the earth's pull ever tugged at the soles of her feet.

We reached the far edge of the fields and entered the woods, and came upon a wooden bridge over the swift stream, a railway bridge on an abandoned right-of-way.

—Oh god oh god oh god, Sylvia said in a kind of happy distress.

The ties and rails had been removed, leaving only flat blacked planking and metal-jointed creosoted members rising to shoulder height on either side. Sylvia walked quickly to the center of the bridge, dropped to her knees, then flat on her stomach, hanging her head over the side. I wondered if she was going to be sick. But then she turned toward us, her hair hanging mussed about her eyes, her face bright, and said:

—My sisters and I used to hang our heads over and watch the trout in the river. When we stayed here while my parents were getting divorced. Well, my mother was in a hospital, and my dad went off sailing, so someone had to take us.

She breathed in deeply, though the scent of creosote baking in the sun was unmissable.

—It smells just the same.

And of course I recalled the cedar closet, and wondered at this odd motif of scented wood in Sylvia's history.

—We pretended we were playing hooky from school. We did miss a lot of school, waiting for someone to come and take us back, and they thought throwing us in a new school out here would traumatize us too much. So if we saw someone down here, a fisherman, or just some kids, we'd dash back into the fields, and hide until they'd gone. Rachel and I knew we were playing, but Alison would get really scared. When I grew up what I mostly wanted to be was truant.

Carla and I had joined Sylvia on the bridge. The river ran shallow over rocks, spoke sweetly. This was the place for the picnic, then, and we settled on the grungy boards, unwrapped the sandwiches, peeled three beers from the six-pack and dropped the rest in still water among rocks to keep them cold. The sandwiches were awful, but we were hungry. No fish appeared, though we dropped potato chip crumbs to entice any that might be about.

I ate half my sandwich sitting on the bridge, then got up and took the other stale half with me down to the riverbank. It was too hot in the sun on the bridge, and the smell of creosote was choking. The beer tasted weak and metallic. My sandwich held something like turkey. The sound of Sylvia and Carla's voices was grating. I tried to listen to the river only, and the wind in the fledgling leaves. My brain was bugged, was trying to work something out, something to do with the well-driller, at least in part. It was not so much being bawled out for something we were only half-guilty of, if that, as it was the man himself, with his ready censure and bedrock bigotry. Why should he make me feel ashamed and afraid? He was the purest form of that strain called grown-ups, and by him you would say it was a mutant breed. How did one come to be that way—stony, angry, arrogant, oblivious? Oh, no doubt he was good to his family, cherished by his friends. So his tombstone might attest when it pressed over him flush to the ground. I knew all adults were not that way, and that most who appeared as he did just wore the clothes and painted on the face. But in time, didn't the clothes shrink and cling, the paint harden and seal? If it wasn't a yellow sport shirt and gray pants, then it was a blue suit and tie. But you must fit somewhere, mustn't you? You must play some role. Couldn't we admit it was all make-believe, and try switching parts, before it was too late? I saw the familiar features of our race in the well-driller's outward shape, but I knew that his heart pumped an alien blood. Or mine did. We seemed to have risen from common stock as different species, that's all. And if he knew what was in my heart, he would hate me. I wanted a fight. I wanted him to fear me. I wanted him to think I was dangerous and subversive to his world where the other was always the enemy. He called us *kids.* He knew we were not from his world, didn't know quite what we were. Then did he think we were dangerous? More likely, he considered us iffy.

And maybe I wondered, on this day of remembrance, why he was alive and my father was dead.

And then there was Carla, planting seeds of doubt—seeds which, I admit, I was tending with compulsive care. Was she content to be Sylvia's friend? Sometimes it didn't seem that way. How they laughed together in the car. Dear god, is laughter a crime?

This kind of thinking was no balm on my brain. I could hear Sylvia talking, her voice a blurting lilt as it found gaps in the curtaining sound of river and wind. She and Carla still sat there baking in the heat and noisome scent of Sylvia's past. She was talking talking talking about things that had happened happened happened. Was that all anyone could talk about, everything that had fucking *happened?* Suddenly I hated everyone and everything. I wanted everyone dead, and that would be the end of the past. Then would I remain, and examine my toes for eternity? I couldn't be sure that the past was dead unless I were dead myself.

Oh what did I care? The wind blew away the voice, pushed it into the river where the cold water took it and the current shredded it over rocks, and all that remained was a greasily shimmering iridescence such as is left by crumbs of potato chips, offered to trout that do not rise. I watched the water. I wished for a fish to rise in a silver sideturned flash to take a crumb or worm or fly that has just nicked the river's skin.

In this way the river gained my thoughts, by bits and fish. A river is that which rives. A river is that which comes between. But also a river is itself, is that which *is* between and is. And in some wise from *river* rose *rivals,* who are two that fish the same stream. From opposite banks. And who eye each other sidewise through cocked reptilian eyes, as herons or egrets, coldly. I caught Carla demolishing a dry crust of white bread, littering the silver sky-faced river. And still nothing rose, and I cast my eye beneath water, where I could. She faced me, and she saw Sylvia, who, oddly, looked upstream as she roiled downstream waters, kicked up silt, dared drowsed fish to strike or hide.

Enough of that then, and I returned to the bridge, pleased with the disruptive thump of my feet on the planked bridge, busting the smelly black intimacy there.

Sylvia was telling of her exile on the farm, her days as orphan and truant. Carla was listening with rapt attention which I found somewhat cloying, for which I rather despised her. They seemed to sense my mood, and only Carla raised her face to offer a slight questioning smile, which I slightly returned without an answer. A sweet, round face, smooth cheeks of pink and olive depths, fine lashes rimming clear blue eyes. Honest eyes, I would have said. Sylvia went on talking talking about things that had happened. Happened.

—August ended, time for school, but we went on as we had all summer, hiding in the cornfields, playing by the river. No parents, no school. It was so good, I never really trusted it.

She was twelve then, Rachel nine, Alison just four. Happy little farmgirl, she helped with token chores, spent the rest of the time wandering on the property, often with her sisters, sometimes alone. It was an Eden, an ease she'd never known. Why did Sylvia hate school so much, if she was so good at it? She was a persecuted child. She was smart, and shy, but even then surely proud and independent, and a bit superior. And neighborhood gossip about her parents' marital problems, in those suburban halcyon days when divorce was still a dark novelty, trickled down as fuel to juvenile abuse. No one, of course, can be as cruel as kids. One of her classmates took up a petition against her. In the forthrightness of its malice it was almost charming: "Sign here if you hate Sylvia," it read. And it garnered an alarming number of signatures, but the girl who instigated it was a bully, skilled in coercion, and, perhaps not coincidentally, the second smartest girl in the class. The teacher uncovered that vicious cabal, and sent home stern notes to the parents of all the signatories, and Sylvia reaped the blame for that, too. And later, when the teacher, following some trend of liberated pedagogy, allowed the class to arrange their desks as they chose, Sylvia positioned herself in the corner behind the upright piano. The other children would peek over the top of the piano at her, or creep around the side for a look. Then maybe call out some abuse, and then flee. Sylvia took it all in fierce silence. Terrible Sylvia. What a weirdo. She was despised. She was feared. She must have enjoyed it, just a little.

But now her father was off sailing a warm blue sea, while her mother, in the cool of some green ward, searched for her scattered wits. The divorce had been messy, with infidelity on both sides, it seemed, and in her father's corner booze and a bad bad temper. He was a big man, blond-bearded, seemingly gentle—Sylvia recalled for us how he used to call her Punkin and help her with her math problems, not that she much needed help, but they both loved mathematics. But he had within him his race's rage of so many winters pent. A kind of cabin fever in the blood. He was a genius, maybe, a math whiz, the first of his family to go to college, and he'd wound up an accountant for many reasons, not least that Sylvia had come along unexpectedly, necessitating rapid marriage, a stable home and income. What we do for those we love, and what we don't give back for it, in time. When I thought of him I pictured a great muscular pink fist, wrapped murderously around a pencil ready to snap. Her mother, Irene, had for her part once considered becoming a nun, had received, in the cedar closet, was it?, a vision of the Virgin. But she was an iron pragmatist at heart, murderously literal-minded, and was perhaps relieved to have her vocation thus silenced. Started off this way, what good could follow? None did. When the divorce was final her father was gone, barred from the house. And off to the Islands to dry out, become a patient man.

—I used to go into the fields and hide there. The stalks seemed so, so tall then, like a hundred feet high, like a cathedral. I would take off my shoes and socks and dig my toes into the soil, it was so thick, and cool and damp like clay, in all that green shade. And I would bunch down as small as I could, and shut my eyes tight, and hold my arms over my head, pushing down. I imagined that I could dissolve into the earth. I knew how plants took up food through their roots. I wanted to dissolve into the earth and be sucked up into the corn. Just disappear. Just gone, like that, so beautiful. I pictured this whole thing, me in molecules, seeping into the ground like water, down, down, then into the root hairs, roots, and up up up into the light, driving up through the stalk, erupting through the ears as silk. Then, I don't know, into the atmosphere, maybe. I must have seen some illustration of plant hydraulics that impressed me. I imagined all this with big red arrows showing where and how I went. I had a guardian angel then, and she was corn, too. Tall and thin. She waved in the wind with slim arms draping out like leaves. She had cornsilk for hair. A little scary, really. I could scare myself out there. I loved it.

Please witness this: It is late September, such a day as makes us willing to endure our home's bitter winters. The corn in the fields stands nearly dry in the soft cool air. Some of the corn has already been cut, and these stubble patches glow in late afternoon sun, and they make you think of the cropped head of a martyr, roughly shorn with a knife or dull scissors, but this silvery orange light, angled by a shelf of gray mackerel-scale clouds edging in from the north, convinces you that this humiliation is suffered gladly, and all is redeemed: the ends are just, and incorruptibly fine. Across the fields, the leaves of the riverbank trees color and shrink as they dry, revealing the trees' perfect, illogical structure, the reaching trunk and noble limbs and exquisitely arching branches: here is *tree* truly seen. This season is the end of adornment. The earth restates its elements. When you breathe you are aware of air; seeing, sky and light assail you; walking, you cannot ignore earth, hard and dry, and your steps raise dust which, caught in the pitiless light, seems like vapor off a block of some caustic element, or as if the light itself has been drawn into the ground by gravity, and flecks off with your steps. The spectrum of light seems to disintegrate, so you cannot tell what colors you are seeing, and must resort to terms like "silvery orange." There are gray and brown, black, silver and blue in one glance at air, and red, orange, and gold, and complexities of fading greens. Oh, it is melancholy, worse, it is heartbreaking, tragic, but—here we are.

All the family are gathered at the farm, an annual picnic, a harvest celebration. All the aunts dote on Sylvia and her sisters, the three small girls engulfed in swells of country kin. The cousins are playing tag, running and jumping like bouncing sacks of feed: that family has the big-bone genes. Sylvia moves among them in a happy haze. With no parents here all is well. Her father's absence is conspicuous and noted, but it feels as if he is beloved, good, and dead, and everyone is getting used to it, beginning to speak of him nostalgically.

They are grilling bratwurst and hamburgers, and roasting corn in the husk (which will be a bit chewy, this late in the year, and not very sweet). An old door, laid across sawhorses and draped in red-checked oilcloth, holds glass and Tupperware bowls of coleslaw, potato salad, and a profusion of Jell-o salads, such that it appears the Stenmarks must have cleaned out the entire local supply of mini-marshmallows and non-dairy whipped topping. There are three dogs, all good-natured mutts, which divide their time between harassing the several barn cats and sitting pleadingly by the grill. Sylvia, Rachel, and Alison are served and seated first at one of the picnic tables trucked in for the day. From her seat on the picnic bench Alison is at eye level with her gargantuan plate of food, and with the late, oversized corn ears, the bulging brats, the unstinting heaps of salad, she seems to have wandered into some giant-sized landscape. They are all three delighted.

Sylvia pulls up the drooping sleeves of her borrowed navy cardigan and they immediately begin to glide again down her thin white wrists. Likewise, the uncles' trousers and straining leather belts cannot contend with the downward push of harvest-bulged bellies. The biggest of the barn cats swaggers by, scornful of the dim-witted dogs, and he too carries a swaying paunch between his hind legs. In the withering market garden the younger cousins are playing king-of-the-hill on a big hubbard squash, which gravity has pulled to a rounded mountain shape, its broad base sagging to the earth like something semi-liquid and still falling. The cousins leap and tussle and fall out of sight in the drying tangle of leaf and vine, then reappear as if born that instant of the earth, springing up with arms outspread, as if they would soar, but down again they tumble. Alison drops a big spoonful of sweet white salad in her lap. Suddenly Sylvia is seeing gravity in everything, the central fact of life on this planet, and life as the always-lost struggle against it. On the swingset the children soar up and up, and down they come, so much faster. The uncles have succumbed at the belly, and the aunts are falling everywhere—wattled chins and upper arms, drooping buttocks and bosoms, thighs sliding down their legs like fat babies on shinny-poles. God's best angel. The Tower of Babel. Every stalk of corn out there will fall. A child appears in the hayloft door, dangling a kitten by its hind legs, teeters on the edge. Someone screams, and Sylvia's breath catches in fear, and then cousin and kitten abruptly vanish, yanked inside to safety. Did her vision produce that panicky moment, the verge of another demonstration? And did she halt it? Sylvia feels omniscient and dangerous. And like a god she feels far away from everything—the noise and banter around the grill, the scurrying doting aunts, the chaos in the garden, even her sisters—but intimate to it, still. In this moment Sylvia, twelve years old, feels dizzy, giddied, with contentment and sadness and an understanding she doesn't understand. She doesn't feel afraid. She notices that the anxious knot she has always carried in her chest is gone. She reaches over to help Alison wipe the salad off her dress. She butters Rachel's corn.

—Come and get it, one of the teenaged cousins yells.

The children quit the garden, flush from the barn, throng about the grill and the bending salad table. A car pulls into the drive.

Everyone turns to look, because no other guests are expected. Sylvia turns to look, and sees the car, with the angled sun on the windshield, obscuring the features of the driver, with dust rising fine and golden around the tires and drifting only a few feet before settling. Whoever is in there sits while everyone watches. Sylvia is both certain and confused, thrilled and horrified. Because she knows the car, and everyone knows who is in there. He is the dear beloved dead one. Dear Dad. Dead Dad. But, oh yeah, he isn't dead, because here he is, and dead men don't drive cars. Sylvia knows not what to think, and isn't thinking. She is responding viscerally, like an animal. This is the moment she has dreaded most, and most desired. Because she loves her father, and hates her parents. Is he ever going to get out of that car? He is savoring the tragedy and beauty of the moment, recalling the hell they have all been through, and imagining the good times ahead, because he has changed, and they do all love each other, they *are* a family. Everyone wishes he would get out of the damned car.

He gets out of the car, looms beside it, adjusts his trousers. He is big, but trim, lean, in fact, from weeks of hard sailing, and he has quit drinking, and become a vegetarian, to spite himself, because he has always been a famous carnivore. He is brimming with his big surprise, with a grin of his big fine teeth beaming through his trimmed beard, and his sea-bleached hair falls across his eyes and he reaches up to push it back, like a boy. He is the baby of the family, but also he is Dad, so it's all a tad confusing. Prodigal father returns. They rejoice, and give thanks, and they slay a fatted bratwurst.

—Will you look at this! an uncle cries.

—Well, I'll be . . . , echoes auntie.

—Come on over here and get a beer, Teddie, someone says, reflexively, forgetting.

Now Alison and Rachel have slid off the picnic bench and they run toward their father. Sylvia sits a moment, then climbs down too, and stands stone still with the oddest look on her face, as love and fear do battle in her brain. And then the battle is over. To the astonishment of all assembled, Sylvia turns and runs for the corn.

There is stunned silence, then puzzled exclamations, then a posse of cousins and the three dogs gives chase. Sylvia has reached the field's stubble border and stumbles bare-legged, in the too-big sweater, toward the rows of dry stalks. There is glee and bloodlust in the cousins' cries and the barking of the dogs, and then disappointed groans, cocked heads and puzzled eyes when Ted calls them off. He joins the watching crowd as Sylvia disappears into the corn.

—It's a shock, I guess. I suppose I should have called.

He holds Alison in his arms and Rachel by the hand. They stare at the field. The top of each stalk has been bent over by the wind, all bent the same way, so it looks like a thin brown crowd waving dreary brown pennants. Hooray.

—Can you beat that, says an aunt.

—What's got into her? asks another.

Sylvia is long gone from sight, but they all stand looking, until someone notices the meat burning on the grill, flames engulfing the burgers and brats. Happily the men rally to douse the flames, and continue with the picnic. Ted squeezes Alison, then sets her down. He tousles Rachel's dark hair. He says:

—Just give her a little time. She'll be back soon enough. She's always been a bit dramatic.

And indeed, once you've run away into a cornfield, what then? Sylvia waits just over the rise, expecting to hear the clatter of pursuit through the dry stalks. But no one comes, and Sylvia's pounding heart settles, her thoughts begin to clear, though the thought that her father is back there in the yard still muddles her. In this place there are no parents. Here the parents are far away, and with them the threat of strife and terror. Still, that *was* her father, and that fact forces her to accept that the farm is just the place where her uncle lives, less than an hour south of the city, a few minutes from the freeway, not a haven, no Eden. They could have her back at *school* in forty-five minutes. Well, she always knew that, though in these blissful orphan weeks her imaginary haven has thrived, grown to encompass all time, displacing all that came before and whatever might lie ahead. But for all that, she is not a dreamy girl, stands a step back even in the most elaborate of her imaginings, knows she is imagining. More likely she would lose herself in math. So she finds herself in a cornfield, and there is not much to do in a cornfield. If they came for her now to take her back, she would go. But can she return on her own? That would be too great a blow to pride, and if they don't even care enough to come after her, then she will stay where she is. She continues along toward the river, dark under the encroaching clouds, sits a while on the bridge. Brown and yellow leaves drift and swirl as the current takes them. Through the dark limbs overhead she sees a coming winter sky.

While in the yard the picnic continues. The cousins career through the forest of adult legs in a frantic game of tag, and no one is sure who is "it." Ted is the centerpiece of the gathering now. He tells about his weeks in the Islands. He relates a harrowing tale of three days in a tropical storm, anchored to the wind, losing the anchor, watching the aluminum mast bend in half like a paper clip, bailing for his life; and his siblings and siblings-in-law, who may have traveled to Iowa, Duluth, or the Wisconsin Dells (wait: Marg and Harold went to Las Vegas last winter), listen with unfeigned amazement and perhaps do not believe a word he says. Rachel and Alison inquire from time to time where their sister has gone, and are told she'll be back soon.

—But it's getting dark, Rachel says.

Merely an appropriate observation, and then Rachel goes to join the game of tag.

Sylvia sees the dark clouds, edged with icy white, sliding across the clear sky. She is back in the corn now, at the top of the rise, and moving from row to row she observes the family gathering in slices. Here are the aunts in their sweaters like Sylvia's, or borrowed windbreakers, seated around a picnic table. They lean plumply back, cross their legs at the ankles, and at arm's length hold cups of the weak coffee, black. Here are the cousins gathered around the grill, poking at the coals with sticks. The wind has come up and it thrashes the dry corn, it fills Sylvia's head with rattle and roar. A general thunder, and specifics of near crisp twistings. Moving down a row, and here are the teenaged girl cousins, sitting and leaning on the hood of a big beige car. They quietly collude with mirrors and combs. The ears are full and dry, the husks peel back and the exposed silk has blackened, like a beard around the deep orange of the hard uneven kernels like crooked yellow teeth. Her father has a beard, not black, and teeth, not yellow, splendid teeth, all thirty-two from incisors to wisdoms intact and flawless, the pride of his mouth. Now blue sky is only a horizon slice. One small white triangular cloud, upturning, sails ahead of the gray sheaf, a spinnaker flying, a chip of white ice. From a juncture above ground each stalk sends down a ring of secondary roots—creepy. Moving down two rows, and there is Father, in a lawn chair, and Alison in his lap, and Rachel leaning against his arm, pulling, hugging, fidgeting. He's talking to someone but Sylvia can't see whom. Now Sylvia is the one who is long gone, fondly, dimly recalled. The girl who vanished in the corn. The roar of the wind on the corn is making her dizzy. The biggest barn tabby enters stage left, and Rachel spins and exits. If the wind made no sound would it frighten or chill or stir us? The winds makes no sound, but things in wind make sound. Wind gives them voice. Inspires them. Sylvia darts back and forth through the rows of corn, trying to see the whole picture, but she only gets these ragged slices, one at a time, like a cartoon crudely drawn. Father—cat—aunts—dog—garden—cousins—car—barn. All of them tiny and distant. All smothered by the crashing corn, waves of a furious ocean, flames of a terrible fire. A freight train, a damburst. She can't look anymore. She turns and runs behind the rise, where she can't see them, and crouches down facing the river. She puts her hands over her ears. She tries to imagine her guardian angel, old silky-hair, protecting her, but the black rotting silk, the crooked toothy kernels, are all she can see, and her guardian angel becomes a horrid grinning witch. The river is on one side, her family on the other. She crouches in the corn and shivers.

Day is gone. The clouds, the brief autumn twilight, the unseasonably warm day have fooled everyone. Now Ted and the uncles gather at the edge of the yard and discuss this situation with Sylvia. They seem to avoid stepping into the stubble verge, as if it were a no-man's-land and dangerous. How fast the dark comes down. How quickly the frail warmth flees. Cousins are dispatched for flashlights. Ted takes the lead and moves toward the corn, calling:

—Sylvie! Punkin! Sylvie? Come on out, now. It's getting dark, honey, it's getting cold. Come on out, Punkin. It's all right. You're not in trouble.

Sylvia hears him. Sylvia knows that it's dark and cold. Sylvia knows she *is* in trouble. Maybe not from Dad, but deeper trouble than that. Sylvia knows that she can never go back. She has slept in the corn, just fell asleep in a crouch, and awakened now to the dark and the cold, and the sound of her name, from the fields all around, her name rising on all sides as if *Sylvia!* were immanent in all things. They're probably not trying to confuse her, but in the dark, in the trackless field, with her name coming from everywhere, Sylvia is utterly disoriented, and wouldn't know where to go if she wanted to turn herself in. She sees lights swinging through the screen of cornstalks. She sees lights pass, dimly, on either side. She realizes they will never find her if she does not want to be found. In the dark the cornfield is a funhouse, a maze. Sylvia sees the lights and hears the voices, but the searchers' vision is closed on either side by the stalks, and Sylvia makes no sound. And it seems now to Sylvia that it is the essence of *Sylvia!* to be unfound. The searchers look into the light; their eyes will never own the darkness where Sylvia hides. As the lights approach she has only to slip down a row or two, and they pass right by, shouting, stumbling. She is exhilarated, incredulous, invisible.

She dozes, and starts at voices, a crashing in the corn close by. Uncle Gene and two big cousins pass so close she can hear their hard breath. They are returning toward the farmhouse with their lights off. One of the cousins says:

—She could have gone across the river into the woods. She could have fallen in from the trestle. The little brat. Shit. It's fuckin' cold.

—Here now, Uncle Gene says sharply. I won't hear that kind of language. She's just a little girl.

The river. The deep dark woods. The hills that plunge east to the sheer fall of bluff over the wide, islanded Mississippi. Sylvia almost panics, in the dark, as the river and the woods and the thought of falling rush her brain. But because she is *Sylvia!* she will fulfill it. She will do what is required for *Sylvia!* to be. Much depends on this. The dark field all full of men shining lights, calling her name—such a lovely thing—relies on her resolve.

She has fallen asleep in front of the television, and Dad is carrying her upstairs to bed. She feels her weight in the big arms circling her. Each step takes them up, up, up; by the top of the stairs they will have gained the momentum, and the will to lightness, which will send them bursting through the ceiling, the roof, on up to the heavens in giant steps, and show that gravity is a fatalist's fiction. Dad's legs are so long, he vaults stars, skips moons. His great feet find traction in the nothing of space. How does he do that? He is Dad.

Three aunts have waited through the night, and they follow Ted up the stairs as he carries muddy cold Sylvia. The water is running in the big porcelain tub. They easily maneuver her torpid limbs to undress her, and lower her into the warm water. She wakes enough to convince them she isn't dead or in shock, mutters something a groggy child would say, so they know she is all right. But her limbs are so cold—it has dropped to near freezing outside. It is four in the morning, and everyone but these aunts has gone home. The men laughed and slapped backs and shook hands as they went to their cars and trucks. No one said what they all must have thought as the night grew colder and their misted breath thickened: they probably found her just in time.

They lift her from the water and wrap her in towels. The rubbing rouses her a little, and with sleepy fascination she watches the last of the bathwater swirl down the drain, leaving a thin wash of silt on the old pitted porcelain. Then into the crisp sheets, and comforters loaded on top. Pale first light through the curtains; her sisters together asleep. The rustle of bed linen is the roar of the corn. She sleeps on the cold black earth, but invisibility blankets her. She has been found, but while she sleeps she still is *Sylvia!*. Voices calling, from everywhere. Whose voice is that? High-pitched, ringing. When she wakes and all is changed, still that voice will remain.

Cheerful enough, that voice, disinterested. She threw things away, her deepest secrets. What good are these to me? She had told me things about her bad times at the University, stories all covered in grime and sickness, dirty clothes, dirty hair, and despair. Brief affairs she had had with men she didn't much like. She had attracted a certain kind of fairly intelligent, ambitious, somewhat defective man. If she had been stronger, if she had cared, about anything, she wouldn't have let them near her, and they wouldn't have dared, but in her debilitated state she was fair game. She told me she had let them make love to her because she couldn't be bothered to say why not. They imagined they possessed her, but Sylvia knew otherwise. They did not own her, did not know her at all, because she had withdrawn to an inviolate center, like a wintering tree, and let all unnecessary extermities fall away like withered leaves and sapless limbs. She let go of everything in order to save the only thing.

But can you tell your deepest secrets? Are they even yours to tell?

And where would we find all that Sylvia had thrown away? In the river's dark swift script? We all looked to it, as if we could find these things written there. I looked there, but I found I was reading another book, the one that told of our favorite saints. I had written a chapter for tiny Saint Sylvia, "Our Lady of the Corn."

—The next day Dad told us that Mom was out of the hospital—he said, "Your mother is better now," fat chance—and that we three would be going back to live with her, going back that very day. I heard it through a deep fog. I was in a daze for weeks, I think. Utter dreamland. Here we are, back in suburbia, here we go off to school, dressed in bright new school clothes. I was a zombie.

—It's hard to imagine you as a zombie, Carla said.

—In some ways I feel like I never got out of the corn. I just learned to adjust to real life as a zombie from the corn.

They laughed. I smiled, I guess. But the joke seemed maybe too true. Not that Sylvia was zombie-ish in any way, but I'd worked so hard to imagine her into this world as real and human, not some fantastic changeling. I'd like you to meet my girlfriend, the Zombie from the Corn. But maybe Sylvia was on to something. She always meant what she said, though she wasn't always serious. We wear masks like devious maps to the hidden self. Looking at a map, you might fall in love with perfectly wretched lands. If you traveled there you might find the people dispirited, surly, oppressed, the food disgusting. I briefly wondered what faculty might stand in as travel agent to the soul, doling out full-color brochures on the psyche, but then my mind wandered. Something upstream caught my eye, something large and brown. At first I thought it was a deer crossing the river, but then it came into sight as a fisherman wading the stream.

—I don't know, Sylvie, Carla said. It's all so *gothic.* Northern gothic.

And she reached out and touched Sylvia's arm. *Sylvie.* With that simple nickname it seemed that Carla had ventured an intimacy I couldn't even have imagined. I didn't know if anyone had ever called her Sylvie, though I had imagined that her father did. I wasn't angry or anxious or suspicious now; I gazed upstream in simple stunned awe at Carla's human virtuosity. Go ahead, take her—you win. I could spin dramas from the scraps of Sylvia's past, mine the beauty and the tragedy, but you, you called her by that name, and touched her arm. You look at her that way, and you smile. I could give to her my imagination entire, but I wondered if I could give her my heart.

The afternoon was wearing on, and I was running down. I leaned on the bridge railing and watched the fisherman as he made his way downstream. The patience and purpose in his movements involved my diminished attention completely, and I watched dazedly, trying to see what he was up to, what he saw on the water. It was like watching a master at a game when you don't know the rules. I only knew about trout fishing from reading, the big two-hearted romance of it, but this man seemed to work the stream with perfect ease and harmony. He paused and looked, considered, looking for something I couldn't see, and when he saw it he drew the long rod sharply back, roping the line out tight behind him, and then shot it forward, and the fly, invisible to me, dropped softly on the water, leader and line falling behind it, slyly. He let the fly drift a few seconds as he bent forward with tingling concentration, twitched the rod tip so slightly. This consuming attention was focused on something that mightn't even be there. He seemed sure that something was there. I sensed faith and knowing in his posture. But he wasn't catching any fish. He didn't seem to mind. He moved downstream in a slow erratic pattern, his reading of the river shaping his path. He was quite near us when a cast to the shore produced a strike, and he jerked back the rod, and a small fish sailed a few feet toward him and dropped back into the water. His body relaxed into a pose of casual disgust. He pulled in line and brought up his catch, flopping silver, no more than finger-length. He grabbed it in his fist, and freed the hook. He held it a moment more in the tight fist, then opened his hand and let the fish drop. It drifted lifeless past me, under the bridge.

By now Carla and Sylvia were watching too. The fisherman reeled in his line and waded toward us. I called hello, and he said:

—Good day to be on the river.

He was just a few feet from the bridge. He could have rested his arms on it if he were closer. Carla said:

—Why did you kill that fish?

—Oh, it's just a chub. Garbage fish. Now he's mink food. Haven't raised a trout all day.

He was wearing a brown felt hat stuck all over with needle-like things which appeared to be porcupine quills. He had a trimmed brown beard and wore wraparound reflecting sunglasses. His waders fit tight; they looked like wet-suit overalls. Under his tattered vest of many pockets he wore a red chammy shirt. A sheepskin patch on the vest held a half-dozen flies.

—It's too clear, too bright, he said. They don't feel safe.

—Are there trout in here? Carla asked.

—Oh, for sure. Probably a hundred in this stretch of water we can see.

—No, I said.

—Absolutely. Not big ones, but trout. Browns and brookies.

I saw him glance toward the three bright beer cans cooling by the shore. I said:

—Have a beer?

He gave an emphatic nod, went over and lifted the three connected cans, pulled one off and held out the remaining two to me. I took them. He opened his and drank, and said:

—What are you folks up to? Not fishing.

—No, I said. Picnic.

He nodded, looking down the bright river.

—I should have stuck to picnicking. But, the river is the river.

I wasn't sure if that was deep or stupid, or maybe just a sort of fisherman's *non sequitur* in-joke. And I didn't get any more evidence for my decision, because he thanked us for the beer, wished us happy picnicking, and ducked to clear the bridge as he continued downstream. I looked with some longing as he waded in the sun through the rocks, sipping at his can of beer.

—Is the river the river? Carla said.

—I would say, it is, said Sylvia.

. . .

We were all a bit weary as we walked back toward the car. The straight path back led us to a plowed field, and, foolishly, we started in across it instead of walking around to find level footing. We teetered and stumbled and bumped like drunks as we tried to navigate the devious ground. The big cleaved chunks were bent on twisting ankles. They collapsed irregularly when we stepped on them, throwing us off balance. About halfway across, we had all swerved together in a triangle with Sylvia in the lead. We bounced off each other for balance, hopeless. Sylvia caught her toe under a clod and pitched forward. Carla grabbed for her, I grabbed for her. We tangled arms and legs and all three went down, Carla landing on the backs of Sylvia's legs, and me on top of Carla. We landed quite softly in the tilled black soil. And no one moved. I was half on my back, looking up at the clear blue sky. And surely some god was looking down on us. Not laughing. Merely calmly considering. I gave him a beard and wraparound glasses. It was actually quite a relief, to be down on the ground after all that furious resistance. I was even fairly comfortable, pillowed by Carla and Sylvia. Finally Sylvia said:

—Ouch.

And we climbed off of her, gained our feet, brushed one another off, and continued on our precarious way to the car.

Drowsiness deluged us in the sun-warmed car. Already it had been a long, long day. And it would be a long drive home, at least an hour, though we could pick up the freeway nearby, cut the Gordian knot of the twisting way Sylvia had brought us. Hardly a word was spoken, except for Sylvia's directing me back to the main road. By the time we reached the freeway,

Sylvia and Carla were asleep, leaning together, Sylvia's head on Carla's shoulder. I drove gently so as not to disturb them, in spite of other impulses. The radio poured forth its indefatigable pop. I tuned around looking for something more palatable. I heard Neil Young assure me that *"the motel of lost companions waits with heated pool—and bar,"* and then I turned it off. I smoked a cigarette that tasted harsh and metallic in my dry mouth. The sun hummed down and the wheels beat round, and I thought we would never get home.

About halfway there Sylvia woke. She rubbed her eyes and yawned, smiled and kissed my cheek. She leaned lightly against my shoulder and played with the hair on my neck. That was lovely, and I smiled, but I felt really quite sad. Carla woke as we crossed the Minnesota again. She was refreshed and cheerful. She had a great idea: she wanted to take us to dinner at her restaurant. Sylvia's face brightened, but I looked at her as I said, yawning:

—Sounds nice, but I don't think so. I'm pretty wiped out.

And Sylvia said:

—I'll take a raincheck.

Which Carla took in stride, with a smiling shrug.

We dropped Carla at her building, declining another offer, this time for drinks. I persuaded Sylvia to come to my apartment for the night, so we could cook supper, so I could move about freely, use the bathroom without fear.

That night when we made love I was run through with a desperation which Sylvia must have sensed, though I am sure she did not understand it. I wanted to be sure that she did not draw away from me, to any place within, no matter how lovely and safe that place might be. What Sylvia said, when we were done, was:

—Whew.

And then I lay on top of her, and held her bare shoulders tight, and spoke to her with an intensity which probably would have frightened her, if she had not been Sylvia. I told her that I loved her and would not let her slip away from me. I said I would force her to stay in this, stay with me, I would use all my will to ensure this. And I told myself that my will was sufficient to this task, to loving Sylvia and making her love me. It wasn't arrogance or ego. I knew she did love me. But, she might just decide it wasn't worth it. Not everyone values love so very highly. Sylvia said:

—I do love you.

I searched the tones of her voice. If she had said that in her dramatic voice, I think I would have killed us both. But this was a calm low voice, not her usual voice, not the dramatic voice, yet another. I looked at her for some moments, as she looked at me. And then I turned off the light.

I did not fall asleep right away. I lay half-conscious while highways and hallways and stairways, rows of gravestones and alleys of corn, and bridges and rivers careered through my brain. Dear dead dads. *The river is the river*—it was a parable or a mantra to my sleepy mind pouring out these impressions. A well-driller and a fisherman, they were like a devil and an angel who perched on your shoulders. And finally the main street of Red Earth, with the old soldiers setting flags along the curb. The wind came up, and stiffened and swirled, raising whirlwinds of a winter's worth of sand and salt and trash. And the wind rose to gale, a hard hot breath from tireless lungs, and snatched the flags the old soldiers had planted and sent them flying tip over tail down the empty main street. Everything the wind could lift, down to newspaper boxes, potted plants, stray shoes, it hoisted and sailed, and dropped it all in a jumbled heap at the base of the broad back wall of the discount store that dead-ended the avenue. Children of all colors crept out from basement apartments, and the wind took them too, and flung them down the avenue and landed them with soft thuds in the dump of flags and windblown flotsam. Above them on the concrete wall a mural showed a happy country scene, rolling fields and jaunty white farmhouses, and round green trees; above the fields walked a giant man in a gray business suit and polished black shoes, square-jawed, carrying a briefcase. He treaded on air, and he was angled slightly upward—soon he would be striding in the real blue sky.

Dirty

Dirty clothes, dirty hair, and despair. In Sylvia's second winter at the University. She has settled in to a grimy existence devoid of expectation. The dirty clothes pile up and overspill the basket. The laundromat is two blocks away. It is always bitter cold—is it always bitter cold?—it can't *always* be bitter cold—it seems always bitter cold. Sometimes she gathers resolve and stuffs the laundry sack, and it sits there like a smudged white boulder; she pictures herself hoisting it in her arms, lugging it down the stairs, out into the frightful cold, down the sidewalk; her tiny arms can't reach even half-way around, she looks like some caricature of an enormously pregnant woman; and then the bag has the weight of a boulder, she won't even be able to lift it. This envisioning exhausts her, she dumps the bag out again, decides she can wear these clothes she has on for another couple of days. It takes her two or three runs at it, two or three stuffings and dumpings of the bag, before she gets any wash done. The dirty hair: what is the point of showering just to put on dirty clothes again? And then, she has this odd fear of leaving her room, even to go down the hall to the bathroom. She isn't sure if she thinks something terrible will happen while she is gone, or if she is afraid of missing something—a cherished friend from long ago and her very salvation will call for her, and where is Sylvia? *Washing her bloody hair, is where.* It is all part of this: That Sylvia is afraid to move, afraid to disturb in any way this stable despair which, for now at least, falls within limits her will can circumscribe. It could become worse; she doesn't see how it can get better.

The rooming house air is sharp, musty, dry. The ancient hallway carpets exhale the stale breath of age, dust, and winter. The radiator rages and clatters with the fits of the deranged boiler. Sylvia goes to bed freezing, in long underwear or even sweatshirt and pants, under a pile of cheap thin blankets, everybody's castoffs, free gifts from banks, and wakes sweating and thrashing from desert dreams, having kicked off all the covers, the room like an oven. When she blows her nose in the morning the desiccated membranes give blood. She works her pasty tongue past abrasive cheeks, splintering lips.

—Water, she rasps. Water.

At night the moths come out. It makes Sylvia wonder if she isn't really going nuts, but they are real enough, the moths, small gray-brown, fluttering everywhere, as natural as can be, giving the sense that they are of some rare species that exists only in Sylvia's room—the EPA will bang at the door, seal the room, throw Sylvia out to perish in the acid cold, to save the moths. They flap silently about the light bulb, walk on the ceiling, rest on the walls. They don't seem to be eating her clothes. In the day they disappear. Where do they come from? Where do they go?

She has a black wool wraparound skirt stretched out and tacked over the north windows to cut the draft. It makes her feel she is in a room of mourning, and she feels an urge to cover the mirror, as well. In this context the moths seem symbolic. In some culture they must signify something, something to do with death. She thinks they should be bats, cavorting. Once she did have a bat in the room. She left hoping it would let itself out, and when she came back it appeared to have gone, but she discovered it hid behind "The Last Supper." It was sleepy and she knocked it down with a ruler and threw some dirty clothes on it and flung the dirty clothes bat and all out the door onto the fire escape. The bat climbed out and sat for some time on the dirty clothes, groggy and dismayed, and then, apparently, flew away. It was the last she saw of bats, and where were they when you needed them? Sylvia lost a sock in that episode.

The pictures sent to or from the imaginary friend are down now. The room is adorned as it will be until she leaves it. How will she ever leave it? She imagines a damp stone cell, chosen for abnegation, and she will kneel on the rough floor, bare-legged, and pray till her knees bleed, pray with such fury and devotion that the end will be self-immolation, leaving a dry charred spot on the floor, the shadow of her fervor on the wall, burned in by the white-hot blaze that consumes her. Can she do the same through relentless working of differential equations? She is only sporadically, desperately religious.

She had a run-in over religion with the counselor she has been seeing. She is ashamed when she thinks about it. Her mother, with her great faith in the modern science of psychology, insisted that Sylvia talk to someone. Sylvia, certain that she was beyond or at least impervious to help, began seeing the counselor just to prove her mother wrong. The counselor worked for a Lutheran social service group as an adjunct to the University health service. He was a Lutheran minister and he looked like Sylvia's father; but that was not unusual, as both were of the indigenous type, and half the middle-aged men in the city, the state, bore some resemblance to her father. The counselor was in his mid-forties, large, fair-haired, bearded. He wore bulky sweaters of brown or beige, with reindeer or Christmas tree motifs. Behind thick wire-rimmed glasses his eyes were kind and searching and also somewhat severe. He was restrained, thoughtful, never hasty. He carried a bit of a paunch. When Sylvia first met him she noticed how his big thighs strained at his trousers when he sat or crossed his legs, and the image invaded her mind of her father in a bathing suit, broad pale belly sagging, white thighs plastered with dark hair in unbecoming patterns. The image was familiar, comforting, but in this context also somewhat frightening.

His office was in a rickety wooden building that overlooked the river, one in a row of buildings thrown up after World War II and still known as Temporary Office Building No. 1, No. 2, No. 3, No. 4. The counselor sat with his back to the drafty wood-framed window, behind a large oak desk. Sylvia sat facing him in a padded wooden desk chair, like a student in conference with a professor, which was what the room was really set up for, in its temporary way. There were metal bookshelves around the room, filled patchily with books that didn't belong to anyone, books produced by institutional publishers to fill bookshelves in temporary offices. It depressed Sylvia to see these books that would never be read; but it depressed her more to think of anyone reading them.

The counselor had had some training, but he wasn't especially psychological. His clients were the homesick, the jilted, the rural kids asea in the big city, the woeful scholars who couldn't bear to tell the folks they'd just flunked another quarter. When a certifiable case showed up at his door he passed it along to abler hands. He and Sylvia didn't know quite what to make of each other at first. Sylvia suffered vague terrors and despair. He didn't have a method for this; he liked an identifiable problem, not because he was naive enough to think that getting a new boyfriend or raising one's grade-point average would yield contentment, but it was somewhere to start, a foothold in the world of everyday concerns, which was very much his world. He was a stolid optimist, he said things would get better; Sylvia was a zealous fatalist, she said why should they? Their attraction was perhaps somewhat evangelical on both sides. After meeting him Sylvia was even more convinced that he couldn't help her—how would sitting in a room with one of her uncles cure her malaise? But it wasn't unpleasant, and she liked him, so she kept going. Plus her mother had said she would pay for it; the health service covered a certain number of sessions, which Sylvia was determined to exceed.

She liked the counselor because he seemed so genuinely puzzled by her, and tried so earnestly to understand her. And she liked him because he looked like her father and any number of her uncles, but wasn't. And she liked him because he smoked. That seemed to Sylvia a kind of chink in his stoically cheerful armor, a sign of buried anxieties which she thought maybe she could expose. Sylvia was smoking then too, and when they came to an impasse, which was frequently, they would sit in silence, smoking, and the room would fill with clouds of gray smoke, as if from the exertions of their overburdened brains and straining wills.

The incident which Sylvia came to think of as The Inquisition occurred when she had been seeing the counselor for about two months. The first month had made her feel somewhat better, just having someone to talk to. The next month brought her to face how dismal her situation was. As she saw it: She had no friends, no family, no hope, no home, no clean clothes, no ambition, no desire at all. January darkened down. She seemed especially susceptible to the cold. She went to her classes, sometimes, she went to work in the lab, she stayed in her room. She sat amid moths and mourning cloth. She contemplated spontaneous combustion. Since her emptiness was all she had, she clung fiercely to it, and to all she had lost. She was in a bad mood.

I must consider, she said to the counselor, *that things simply might not* get better.

He nodded as she spoke, but it was the automatic nod of the uncomprehending. That things would get better was his keystone, his hole card, his piece of the cross.

I wrote in my journal: "Each day the bed of gray ice draws me downward."

He wondered if he could suggest she take the bus when she had to cross the river.

The enormous brown glass ashtray on the desk was heaped with the butts of Sylvia's filters, the counselor's unfiltered Pall Malls. The counselor liked the red package.

Sometimes, just walking across campus, I panic, I feel terrified, for no reason, I want to run and hide.

Do you?

Do I what?
Run and hide.
No, of course not.
Well, that's something, isn't it?
What's something?
That you don't run and hide even though you want to.
Why is that something?
It shows you have the strength to resist it.
Why shouldn't I run and hide?
What kind of way is that to live?
What kind of way is wanting to and not?
Wanting to . . . and . . . not . . . ?
Wanting to hide and not hiding, wanting to die and not dying.

An impasse. Gray billows in the column of sun through the window. The river below, a dark channel of water snaking through the ice.

If I died now would I go to hell? If you kill yourself do you go to hell? Do you believe in hell?

It started there. No, the counselor did not believe in hell, nor heaven, nor purgatory. For a religious man he was singularly ambivalent and vague about just what happened to one's immortal soul after death. But as it turned out the counselor, the reverend, was not a particularly religious man. He wasn't entirely sold on the divine nature of Jesus the Christ. It followed that he wasn't too keen on God.

I don't think many people these days take the Bible at literal face value, said the counselor as Sylvia regarded him gaping. *The miracle stories we take to be illustrative and allegorical. There are reasonable scientific explanations now for many of the "miracles."* He etched the quotation marks in the air, about as firmly, it would seem, as the Lord had struck the commandments in stone.

Jesus was an exemplary man and a remarkable teacher. We take the example of his life very seriously. But the virgin birth, and the resurrection, at a factual level, are a little hard to take, aren't they? We see them as symbolic.

We we we, he said. It was a conspiracy. This man was a secular humanist. He masqueraded as a Lutheran pastor, when in fact he was a goddamn Unitarian.

Oh no, Sylvia said.

What's the matter?

Oh, no, said Sylvia. Sylvia had been brought up Catholic, and confirmed with great devotion and pride, and she wasn't sure if she'd been back to church since. Her early indoctrination had succeeded to the extent that she believed that she was and ever would be a worthless sinner. She did not resent this, but cherished it as her last link to a lost body of belief. The church, the Christ, the chance of grace—even if she did not belong to it, it was there, something solid, a source of solace, that encompassing body of faith and forgiveness. She could not bear the thought that the people entrusted to administer it treated it, unashamedly, as a sham, like advertising men who cheerfully admitted that their product was without redeeming value, and went on flacking its bright image just the same.

Then what's the point? Sylvia said. *Why do you keep doing it? How can you lie to people like that? I can't believe I'm hearing this.*

The counselor sat back in his chair. The chair creaked in the way it did only when the counselor recoiled. He lit a cigarette and with the thumb of the cigarette hand cleared stray beard hairs away from the corners of his mouth—another tell, a habit that exposed uncertainty.

We're not lying *to anyone, Sylvia. We believe in the traditions, and in the spirit that unites the Christian community, and the rituals that bring us together. It's a theology of love, trying to do good and be good. It's not so easy anymore. People are confused.*

Don't say theology, Sylvia snapped. *You don't have one. And why is it always* we we we? *What do* you *believe?*

Now the counselor was becoming a bit annoyed and who could blame him. He tipped forward in his chair, and jabbing with his cigarette at Sylvia, said,

Do you believe in the old-fashioned God, Sylvia? Do you believe in a great big bearded fatherly God sitting up on a big chair in the clouds? Then why should I? Do you think that when I was ordained I threw away my powers of reasoning? Just because I'm a minister doesn't mean I'm an idiot.

You should be an idiot, Sylvia said, her head bowed and tears coming to her eyes. *Reason is for shit. What is faith for? Believing in spite of reason, isn't that what faith is for?* Sylvia, preaching as one of the damned, nonetheless spoke with conviction. *What are you trying to pull? It's all a big lie.*

We're not some terrible cult, Sylvia.

Stop that we we we.

The counselor didn't want to make Sylvia cry. He felt a professional and a Christian obligation to her, and he knew he was failing at both. But something compelled him to keep explaining, though he knew he wasn't explaining. Something made him keep talking.

I think you have a pretty outmoded notion of what the church is and does. Our ministry is to bring people together in community, and to help and guide them in difficult times.

Why not just join the Rotary? Sylvia wailed.

We take seriously our role as Christians. We just view the life of Christ, and its modern relevance, differently than you apparently do.

Oh, god, don't say relevance. *Jesus is the Christ and the Son of God, Jesus died for our sins.*

We do believe that, of course, in a sense.

Christ and the Lord forgive and shepherd. Please don't say in a sense.

The counselor hated the tone he heard in his voice as he spoke of the differences between Protestantism and Catholicism, the great mystical tradition in the latter, which no doubt gave Sylvia a somewhat different slant on these matters. That wasn't especially practical in ministering to his flock.

Practical, Sylvia whispered.

The counselor said something else, he wasn't sure what, because in the midst he noticed Sylvia hunched in her chair and weeping, hands over her face, shoulders heaving. He stood and moved around the desk toward her, but before he reached her she was gone. The flimsy Temp. Off. Bldg. door smacked a metal bookshelf and the door thrummed shivering and the bookshelf rang, and the counselor stood in the pale uncomforting winter light with a watery glint rimming his puzzled eyes.

The bookshelf's reverberations faded and ceased. The counselor heard happy talk from a radio in an office beyond the thin walls. The floorboards gave under his foot as he turned back toward his desk. On the bookshelves were propped thank-you cards from students he had helped, cute storebought cards,with clever heartfelt messages. There were letters of the same sentiment in the drawer of the desk. So he was not much for raptures and visions, and perhaps a bit too firmly grounded in this world. Could one be too firmly grounded in this world? Did we have proof of any other? Within the limitations of his faith, he did the best he could. He believed that what he had was truly faith. If it wasn't, could it have been so hard?

As Sylvia remembers that day she finds her mouth moving with the words she spoke, though no sound emerges now. She is staring out the window. Or she thinks she is staring out the window when in fact she is looking into the black cloth that drapes the windows. Because she is in the habit of circling absently in her room and then stopping to stare out the windows, she does the same now even though they are covered. She holds her arms folded tight across her chest, and when she emerges from her trance she drops her arms and shudders. She thinks a long hot bath might do her good, but you can't take a bath undisturbed in the shared bathroom. And she would have to scrub out the accumulated scum. And of course she is also afraid to leave the room. She remembers the counseling sessions with her family after the divorce, and how because of that memory she had dreaded going to see the counselor. But seeing the counselor is different, and usually better. Those earlier sessions were her mother's idea, of course. To heal the family and make all the secrets known. But it all proceeded according to a script Irene had written in advance. Sylvia recalls saying the things she knew she was supposed to say and saying nothing true, or if true only coincidentally, accidentally so. It was all to humor Irene, it seemed everything was. But why? Was she so fragile? or such a threat? Neither, both. So everyone read the lines. The whole family is dramatic, it has to have come from Irene. Irene was so good at it, she never chewed the scenery; Sylvia had a good touch, too, though she sometimes went over the top; Rachel didn't know it was a play; Alison threw herself into it; Richard was limited but effective, like Clint Eastwood. Sylvia thinks that Irene at some level believed it, and still held some hope. She did want things to be better, for the family to be whole and loving, however it was constituted. And she did carry guilt, and so she did try. Sylvia knew that her mother did try, and did care; that was the worst part of it. Sylvia wonders why lately she seems to be attacking everyone's faith and best efforts. She is a crusader in the cause of nihilism. Not that she wants to be, but that is the view from where she stands.

She sneezes, and again, and again. Out goes her soul, and no one there to save it for her. The dust, the must, the moth dust. It takes so little to lose a soul. She solaces herself with math for a while. If science can explain the miracle stories then perhaps faith will pierce her heart through the numbers. She tricks out every proof, without revelation or conversion, then goes to bed.

Next morning when Sylvia wakes she goes to the window and throws back the black. She went to sleep in a grimy sweatsuit and wakes naked, with covers thrown down in the dust. Her bare chest is flushed, her bare arms warm, and she feels waves of cold off the thin cracked glass. Along with the cold, a rush of elation, in waves, a wash of good spirits, first thing in the morning. Simply put, she feels inexplicably happy. She smiles, and her face feels the strain of that unaccustomed shape.

She ponders the small happiness, which to her is not small but like a great weight lifted, a gulp of air to a drowning woman. She knows that the immediate cause is certain electrochemical signals in her brain, triggers and levers responding to precise stimuli, like railroad switches sending through this burst of elation, the 7:22 Small Happiness Express. But why this morning and not another? The sun is just up on a clear day after many days of gray. The light has something to do with it, and probably her dreams bear upon it, even whether she has peed yet. And then maybe it is all the work of some renegade synapse and has nothing to do with anything external, is simply her brain all alone, shooting its mouth off.

No classes today. Lab work in the morning, then lunch with her father, she remembers, and her session with the counselor in the afternoon. She puts on her robe, goes to the bathroom, brings back water which she boils in a hot pot for instant coffee. She peruses the old food in the fridge and decides she will skip breakfast. She always skips breakfast, usually eats lunch, often skips supper. Lately she has grown so thin, even her sturdy thighs have dwindled. She smokes a cigarette with her coffee, considers the day, without dread, for a change. Appointments and responsibilities tend to bring on an anxious sweaty surge, but today she sees them as a way to structure and prolong the small happiness. She is eager to begin it, then, and she showers (!) and dresses, and descends the main stairway through the sounds of morning talk shows and zooming hair dryers, girlish busy-ness in the hallways and bathrooms. Mrs Simic is in her parlor just off the entry hall, sitting in state in an overstuffed chair, plump in her black gabardine as always, listening to farm reports on the radio, loud. Her eyes behind the thickest imaginable glasses do pick out Sylvia as she lands, and Mrs Simic says "Good morning!" with some surprise, since she rarely sees her, knows she is using the fire escape to come and go, which is prohibited. Sylvia returns "Good morning!" with a smile, and Mrs Simic's parting remark is lost in the price of pork bellies as Sylvia goes out the front door.

The campus is quiet with eight o'clock classes just begun. A heavy frost has whitened everything, making delicate and lovely the grim gray trees; the stodgy stone buildings appear translucent in the pale flocking; the icy mist in the air spreads the soft light as if with beneficent intent. Is it warmer this morning than usual, or is happiness proof against the cold? The few students she sees walking quickly across the main quad seem earnest and good-natured. Too soon she has reached the lab complex. She is sorry to leave this splendid light and air, but work, too, is good, and the company of her colleagues, and the care of her charges, the mice.

Sylvia unlocks the lab door, and at the sound of its opening the mice run to the front of their cages, a bank of cages on the wall opposite the door. They twitch and sniff and strain their pink ears, and stare myopically into blurry space, waiting for their master and the instrument of their fate to hove into focus. Sylvia says,

—Hi kids.

And does a unanimous chirr of pleasure, scarcely audible, quiver through the air? Sylvia puts on her white lab coat and dons her radiation badge, and walks across the room, smallish, with a center island bearing sinks and gas jets. She stands in front of the cages, and smiling, hands on hips, says,

—It's a good day to kill some rodents. Hey, Bossie, hi, Petruchio.

Only a few of the mice have names, the long-timers. Sylvia checks a clipboard and then gathers equipment: a white enameled tray, a pencil. She locates first the mouse identified as A-201/DL/07-26-81 and better known as Uncle Joe, because he appears to have a bushy square mustache under his nose.

—Joseph Stalin, time to pay for your crimes, Sylvia says.

She opens Uncle Joe's cage and brings him out, and he escapes her grasp and scampers up her arm, over her shoulder, and twitches at her throat, tickling with his whiskers. Sylvia laughs, retrieves him, gives him a visual check-over. He seems so delighted by her company and attention, and not the least bit afraid. What *does* a mouse think? No thoughts that require more than one-point-five seconds to think, she knows. So they don't remember. Their reaction to her entering the lab, to the sound of her voice, is merely conditioned reflex. So they say. Are they pleasant, these split-second thoughts? The mice are given food and water and a cozy place to nest. From time to time they are allowed to do what mice do best, which is making more mice (and sometimes accidentally, by an error of sexing). They are relieved of all the dangerous duties of a mouse on the loose, never flee in terror from predators, don't know that Sylvia and her colleagues fill that role. But in one-point-five-second dreams did archetypal loomings of broad-winged shadows shrink their hearts and set them pounding? Are these mice *happy?*

Thinking on the emotional life of mice, she remembers a standardized test she once took that included a reading comprehension passage about elephant behavior. It said, among other things, that elephants touch each other with their trunks to express love. Stated just so, a cold hard fact. And in a fit of scientific indignation Sylvia had scrawled a rebuttal in the test book *(you are never, ever to write in the test book)* decrying this absurd, naive anthropomorphizing, this hackneyed sentimentality; Dumbo and Bambi notwithstanding, she had quipped, animals had no emotions, and the passage in question was stupid, stupid, stupid, and wrong. When she had finished the whole margin of the page was filled with her scrawling tirade in No. 2 pencil. Then she went on to answer the questions, pressed for time.

Now the incident floods back on her, and she feels embarrassed and ashamed. She feels bad for both the test people and the elephants. Who was she to presume upon the elephant heart? How could she be so certain that they felt no emotions? Why assume they didn't? Why else did they touch each other like that? Maybe it wasn't human love they felt, but if it was elephant love, couldn't you call it love, just the same? She fantasizes for a moment about seeking out the test people and apologizing. She consoles herself thinking that much of what she wrote, in her tiny hand in pencil on the cheap paper, must have been illegible.

In lieu of elephants she apologizes silently to the mouse. The mouse twitches at her, snuffles in her palm, and if it doesn't forgive at least forgets. But an elephant never forgets. Elephants are afraid of mice, in cartoons. And they bury their dead or something like that; they clatter and mourn among their ancestors' bones. Like Hamlet they understand mortality.

Sylvia is pouring Uncle Joe from one hand into the other when Robert, her supervisor, comes in. He says good morning and remarks that she is in early today.

—I have a busy afternoon, Sylvia says. I have a lot of people to see.

She almost chokes on *people,* the word seems to express all human joy and belonging. The small happiness has rebounded from the elephant incident.

Robert puts on his lab coat and goes about his chores. Sylvia checks the clock to see when she really started working. She looks over the chart again.

—I'm going to be out of town next week, Robert says. Do you want to pick up some hours?

Sylvia lays Uncle Joe on the center island countertop, presses him down spread-eagle.

—I have a lot of class labs next week, but I could use the hours, she says.

She lays the pencil across Uncle Joe's neck where the spinal column meets the skull.

—Take a look at my schedule and see what you can pick up, Robert says.

She takes hold of Uncle Joe's tail near his rump.

—Thanks, she says. I will.

She pulls the tail sharply but with controlled force. At the beginning she pulled the tails off a lot of mice. She lays Uncle Joe on the white enameled tray.

—That was Uncle Joe, she says to Robert.

—Oh no. Joey, we hardly knew ye.

Five more mice die, and go to the blender, the centrifuge,

the expensive machine that counts up cells marked with radioactive dye. Are they mourned, for one-point-five seconds, at least?

Sylvia has finished today's work, with a half hour still before she must go to meet her father. She could get a start on tomorrow's batch. Bossie is on that list. With his black and white patches he looks like a tiny Holstein. Sylvia stands in front of his cage. He is her favorite. He will live another day.

—I meant what I said and I said what I meant, an elephant's faithful one hundred per cent, Sylvia intones as Robert comes back into the lab.

—What?

—Nothing, just talking to the mice.

Sylvia comes up from the lab and into the fine day. It has brightened somewhat, but a sense of morning still prevails. These deep winter days seem to be half morning, half evening, then night. She crosses the avenue to catch a bus downtown, and one arrives just as she does, she climbs on, she finds exact change on the first try, she is charmed.

She came on the bus behind a group of students, among them a clutch of sorority girls. Sylvia wonders for a moment at how sharply she can distinguish the various species of University students, as sure as Darwin picking finches. There are five or six of them, and they wear no hats, thus avoiding the dreaded hat-hair, the flattening, deforming effect a stocking cap would wreak on a moussed and blow-dried 'do. So they must live near enough campus to travel from warmth to warmth without frostbitten ears. They are not going to class on the West Bank, for they carry no books; they carry dress purses—they are going shopping downtown. They take up three double sets of seats, and Sylvia settles just behind them. As they turn to speak to each other Sylvia sees their soft powdered cheeks, their splendid teeth, the sweep of shiny hair behind their clean, cold, red ears, gold-studded. And their voices, when they speak—it is as if they speak some exotic and unbearably beautiful dead language. The content is banal, and the voices are not pretty, are in fact nasal and slightly whining, and full of slurs, mushy vowels, dropped g's—*"Are ya gohn tuh that pardy . . . ? Yeah-r-you?"* But there is something in this speech that is like birdsong, flutes, finely wrought bells. In this perfect speech the sounds are arbitrary and meaningless; the inflections and tones carry the meaning, when there is nothing to be said. It is like audible telepathy, or the strange sympathy twins are said to share. It is the voice of youth and belonging, Sylvia decides, and it is a foreign language to her. More than that, it is a speaking from being, untranslatable to the uninitiated. *Girls, girls, girls,* Sylvia thinks. She would fly into a knee-jerk rage if anyone called her a girl. She doesn't want to be like them, but does she feel a pang of nostalgia for a possibility lost? Wasn't she almost one of them, when she went out with Chad? Correction, Sylvia: You *were* one of them, though only marginally. With whom did you swap lipsticks in the ladies' room of that fabled deco disco? To whom did you confide after that momentous autumn night with Chad? But back then (oh way back when, almost two years ago), they seemed so sophisticated, womanly. They are the same girls, eternally; Sylvia feels old, worn out, but it is all at a pleasant distance, and the small happiness endures.

Sylvia debusses with the sorority girls downtown. They dance away down the slick brick walk toward the department stores, which Sylvia imagines as magical forests where they will discover the talismans and regalia of their clan—sweaters, lipstick, eyeliner, and shoes. She goes the other way, toward her father's building among banks and brokerages.

She enters the building under an art deco bas-relief of Commerce and Industry. She is early and knows this may upset her father, but she feels daring, impetuous. Up she goes in the lovely old elevator handsomely trimmed in smudged brass. Happily she approaches her father's department where the kindly old secretary recognizes her, smiles and greets her, and says:

—But, Sylvia, your father just left to meet you for lunch!

Oh. Happiness is befuddling. Now she remembers that they arranged to meet at a restaurant near campus. The small happiness is headed for the rocks, but Sylvia checks her watch and sees that she still has twenty-five minutes to intercept him.

—He said he had some errands on the way, laundry, health food store, the secretary says.

Sylvia thanks her and dashes from the building. As she runs she considers, plans, decides: a bus through Northeast will bring her directly to the restaurant, but they run infrequently; buses back the way she came run all the time, but leave her on the far side of campus. She will take her destiny in her legs, and she is bus-lucky again, arriving at Fourth Street at the same moment as the 16 bus.

There are no sorority girls on this bus, but inbound commuter students in stocking caps, parkas, and gloves, huddled with their backpacks full of books, lethargic in the stifling bus air, roused momentarily by the inrush of cold at each stop. Sylvia is nearly bouncing in her seat, egging the bus on, praying for uninhabited stops and dreading the *doing!* of the "Stop Requested" signal. As soon as they reach the bridge Sylvia is up and waiting by the rear exit door, which when they stop sighs grudgingly open with maddening pneumatic torpor, stalling midway so that Sylvia nearly leaps smack into it, finally lurching open, discharging Sylvia as if she was kicked from behind.

Then she is running, over the pedestrian bridge, down the

long mall of the main quad. Her legs feel it—she can't remember the last time she ran—but it feels good, she feels light and fast. And she finds that she is thinking of herself comically—silly, absent-minded Sylvia, what are we going to do with her? A ridiculous, lovable creature. Her father holds to strict schedules, and he'll be mad if she leaves him waiting at the restaurant, but she knows her good spirits will quickly win him back.

Near the edge of campus she must slow to a walk, breath rattling in her lungs, the cold air stinging. She's been smoking too much. Her father will smell it on her and act disapproving. So what?

She checks her watch as she enters the restaurant—three minutes late, within the ordinary discrepancy of watches, but her father is already there, checking his watch, too, and as he stands and moves around the table to hug her, says,

—Hi, sweetheart, it's great to see you. Could you try to be on time? I hate to waste my lunch hour sitting here by myself.

—Hi, Dad, sorry, she says, squeezing hard.

—I thought you said you were going to quit smoking.

—I am.

Her father wears a suit of a too big and obvious plaid, and its lapels are too wide, and so is his tie. It's not just that his wardrobe is years out of fashion—even when men were dressing like this, Sylvia thinks, there was something slightly hayseed in her father's appearance. And in her mother's too, she realizes: Irene sees herself as the modern career woman, stylish, smart, but is drawn to gaudy belts and cheap showy earrings. Why can't Mom accessorize? Something of the farm remains in their attire, and, Sylvia thinks, maybe in her own, too. A willful gaucheness, proclaiming their agrarian roots, like the flaw an expert knitter will leave in a sweater as a signature, though this family dress defect was less conscious.

Her father hands her a menu and at the same time motions for the waitress.

—Do you know what you want, Syl?

The waitress approaches and Sylvia wonders if this is the one who comforted her when she broke down weeping here over a cup of coffee last winter for no specific reason. She hasn't been back since.

—I guess I'll have a sandwich.

—What kind of sandwich, honey? her father says, smiling apologetically at the waitress, who is smiling too and in no apparent hurry.

—The Monterey Jack with guacamole is good, the waitress says.

—That's fine, Sylvia says.

—And I'll have the chef's salad, says her father. And we'll have some tortilla chips and salsa to start with.

The waitress reaches for the menus and Sylvia knows this is the one, with her long soft hair, her thin wrists emerging from the wide sleeves of an over-big sweater. Sylvia smiles at her and she returns the smile, and Sylvia isn't sure she remembers, but she thinks she may be in love. The woman held her hand and gave her soothing tea, and kept her at a quiet back table until she was calm enough to go home.

Their table is in the solarium off the back of the restaurant, with a view of the alley, garbage cans, shipping pallets, the impound lot of a towing service. Everything looks pretty in its delicate pale frosting. The tortilla chips arrive and Ted begins to devour them, his flawless thirty-two producing a terrible, impressive roar. Sylvia is hungry too, but she's a little dizzy from the run in the cold air, and the crunching hurts her head. With the sharp corner of a chip she spears pieces of hot chili from the salsa and hoards them in her cheek, and when she has a good store, bites into them. Her mouth burns, eyes water, nose runs. The rest of the food arrives, and Ted forks tremendous clots of raw vegetables into his mouth. Alfalfa sprouts trail from the corners of his mouth like the entrails of tiny animals. Sylvia imagines him lifting mice by the tail and dropping them into his gnashing maw. She smiles.

—This honey-soy dressing is really good, he says. Want a bite?

—No thanks, I've got plenty here, Sylvia says.

Ted is talking about his problems with his girlfriend, Ellen, a genial woman who works in public relations or advertising or decorating. Sylvia always pictures her as smiling and translucent. Her father has been seeing her for years, and they are always on the verge of breaking up. Then it's on to the kids and Irene while Sylvia begins to dismantle her sandwich—it is five inches tall, three inches of that alfalfa sprouts; she scrapes away the sprouts with her fork, scattering them widely over plate and table top. Alison is growing so fast, she's developing problems with her joints—endless doctor's visits, specialists' bills. Corrective shoes might be made of gold.

—She's covered by *three* god damn medical plans, but do you think your mother will help get it straightened out? I swear. . . .

Her father's hand has frozen in a three-fingered claw. Sylvia eats the guacamole with her teaspoon, and considers eating the cheese, but she has no interest in the dense dark bread. The wearying jag of complaint goes on. Sylvia feels calm; there's a brimming bowl of liquid in her belly; the bowl is thinnest bone china, milky white, the liquid is some element lighter than water, almost lighter than air, greeny-blue swirls; the liquid as it swirls rings oh so faintly the bowl's paper-thin rim; it steadies her, directs her with complete assurance.

When she is with her father she always feels as if they have just come through some terrible ordeal. So the air is harrowed, but the danger is past, they are beginning to recover, and be grateful. This calm she feels today, she thinks, is like the calm that came to her that day on the farm, the day her father returned. The pleasant distance, the measured understanding. The brimming bowl gives the possibility of grace. She looks at her father and the bowl fills to its very edge, and above, surface tension bending the liquid over the bowl's rim. She is surely depressed, in some clinical way, and maybe she is mad, but none of that matters right now. What matters is how her father loves her, more than anything on earth, more than any family, more than he ever loved Irene, more than Ellen, more even than Sylvia's sisters, for we must say truth, now if ever. Sylvia is for him the one: the cause of all he loves, and all he has lost, all that he could never have imagined losing, still cannot. His oldest child conceived unintended by parents too young, the cause of an unwise marriage. She is the sign of his enormous love, and of its failure. It isn't easy for him to deal with her, even to be with her. Sylvia knows this. It is a rock in the bottom of the bone china bowl, rough, gray, uneven; it shifts in the bowl and sends a deep muted clangor through the immaculate liquid; it is what raises the liquid to tremble at the bowl's razor rim.

The waitress returns and gazes benignly on the wreckage of Sylvia's sandwich.

—All finished here? she says.

Sylvia says yes, it was very good, and she and the waitress share a smile, a secret communication; in another instant they might burst into wild laughter and jump on the tables, dance, smashing plates with their heels, strip Sylvia's father naked and fling him into a vat of honey-soy dressing. The waitress moves away with the plates.

Her father picks up the check, and Sylvia leaves an extravagant tip beside her father's niggardly one. He pulls on his old brown carcoat, the one he has owned since he was first married. On the sidewalk he reaches for her, saying,

—Okay, Punkin. It's great to see you.

They hug hard. Sylvia pats her father's back with her mittened hands.

—Do you need anything? he asks.

—Just the usual, says Sylvia.

—Okay, I'll see what I can do. Tax time coming up.

He's such a brilliant accountant, tax time is like Christmas. He asks if he can drive her somewhere, and she says no, she's going back to campus. From the car he waves and smiles, and as he pulls into traffic Sylvia sees his face set into its perpetual stony worry.

Sylvia doesn't know whether to tell the counselor about the bowl in her belly, the green-blue liquid. Well, the liquid changes color, sometimes. The blue and green sometimes run in separate currents, or blend to aqua or turquoise. And rarely it goes rosy, and maybe once even gold. She doesn't know if he'll understand. But she owes him, so she tells him. And they talk for some time about the bowl, about her father, all very cool and almost abstract, as if they were talking desultorily about a painting.

But at the very end, as Sylvia is preparing to leave, he can't help himself:

—Might things be getting a little better, Sylvia? he says.

The extraordinary day of human involvement has worn her out. She lies on her bed that evening utterly weary, fatigued in every part, her legs, her arms, her lungs, her ears, her tongue. And her brain, which seems to have shut down all capacities save those needed for school and work and the most essential mechanical motions, feels stretched and strained in unaccustomed ways. Sylvia lies supine on the bed, watches moths overhead. She thinks they are flying reconnaissance missions over her, she thinks they are concerned by her extreme torpor. She is touched.

The weariness in her tongue is odd, she hasn't really spoken that much—with her father it was all listening, or, anyway, hearing. But maybe the small happiness has caused her tongue, as well as her brain, to twist in strange ways. Better. She recalls times from last winter, and even this past fall, when a grave inertia weighed her tongue, and silence pressed around her like heavy cold water. When hours went by, and days, and she did not speak, and she did not speak, her tongue felt thick and awkward, a knotted muscle, and she had to think how to use it, to make the sounds that people took for speech. She thought she could lose the gift of speech entirely and never again utter a coherent sound. Then would she speak a language of howls and cries, just to stir the icy air? Even she would not decipher it. Sometimes a panic took her, and she had to speak or she would drown in the silence, and the effort to work her tongue was precisely the desperate thrashing of a drowning woman's limbs as the water thickened, the tide dragged seaward.

Forced to think so hard on speech, words came to seem unworldly strange. People shared them like a drug. She wondered after motive. Someone said: *But Boscovitch's dictum is just one interpretation of these phenomena.* She could have said that, but why? All through her days she heard people talking, she decoded the utterances, she understood the terms and the tones, casual, technical, authoritative, intimate. But scarcely anything she heard was original or interesting, or uniquely expressive of the speaker's being. Anyone could say those things, so why bother? These commonplace utterances, were they like calisthenics, to keep the tongue toned for the time when real verbal proficiency was needed? Or was it all a desperate attempt at reassurance, the way monkeys (or maybe, elephants?) touched each other continually, to ask, *Are you still there? Are you still there?* And discover, *You are. Then I'm not alone.*

Are you still there?

But Sylvia was convinced she was alone, and no volume of language could tell her otherwise. Still, whatever the motive, and whether futile or not, Sylvia marveled at the proficiency and fluency with which people performed this commerce of tongues. It came to seem like some magic trick, childishly simple to all appearances, rudimentary in its parts, but requiring some knack, utterly simple when you knew it but impossible to guess, and she didn't know it, and couldn't find out. The words of others drew them together, made bonds and bridges, and warm conduits of emotion, almost fleshly palpable. And they hardly seemed to *try,* their words flowed as naturally as a river falling in its bed. Her words (such as they were, these twisted things she forced forth at such effort could hardly be called by the same name as theirs)—her words, when she blustered out an utterance, emerged stiffly, not as a flowing stream, as rusty links of chain, crudely formed and coarse and useless. They hung there in the air, a crude kinked loop of rusty chain. People looked at it, embarrassed for her. She thought they were. She would have been. Or they reacted with reflexive disgust—*Did* you *do that, Sylvia?* And she wanted to pluck the thing down from the air, tuck it under her arm, and flee.

She knew she could not succumb to the nullity of silence, and these words were all she could salvage from that black well. Her thoughts still came more easily, but more and more it seemed that they too were snatched from the void.

. . .

Better? All undone by this day of minimal activity, a few ordinary conversations? But she remembers when to act or converse at all seemed more than she could do. So, better.

There is a light knock on the fire escape door, and it opens inward, lifting the black cloth, bringing in a gulp of cold air. Sylvia is dozing and starts up, turns to see a man emerge from behind the black curtain, like a magician—but if a magician, one who travels with a grade C road show. His hair is long and he wears round glasses and an army fatigue coat. He shuts the door with bare chapped hands. He turns to her.

—Hi, Sylvia. Shit, man, it's cold.

—Why don't you wear a hat?

—I forgot it.

—Where're your gloves?

—Forgot'm.

Sylvia watches him as he takes off his coat, stands chafing his hands by the radiator. She recalls the day's encounters, this the last. Replaying the bus ride with the sorority girls makes her think of Chad—probably those girls know him. And she thinks that if Chad saw her visitor, in his array of thrift-shop clothes, with his impeccable disregard of grooming and propriety, and that wild look in his eyes—she imagines, believes, that he would be hurt and confused, even now. What have you sunk to, Sylvia? This boy is the smartest boy Sylvia has ever met. He can't remember his hat and gloves, but he is tops in his class in both physics and political science, his double major. He is a fervent Marxist. He has been trying to convert Sylvia, but she can't see herself as part of any class, so it is hopeless. He sees himself as part of a class, sees all relevant contexts that define him. His problem, Sylvia thinks, is precisely this, that he is too easily definable by his beliefs and capacities; he lacks imagination, and so he seems to her insubstantial. To his credit, he is sentimental about nature and fishing.

He pushes her legs aside to sit on the bed, and they talk for a while about classes. Mostly he talks and Sylvia listens. He starts off slow and mumbling, and then he becomes excited, talks fast, laughs, gestures with his hands. He's talking about something funny that happened in a physics lab, but Sylvia can't follow it. He really is a nerd. Sylvia is recalling her day again, remembering as much detail as she can, as if to fix it in memory, as if those unremarkable events were something she could not bear losing. She won't tell him about the sorority girls—he would never understand the tenderness she felt for them, the envy. There are no Marxists in Greek houses. And she will never tell him about the brimming bowl. Maybe she underestimates him—she tends to do that with men—but she can't risk it. He could shatter the brimming bowl, without even meaning to. She won't let him near it, not now and if not now never, because she knows he will be gone soon enough, and it is only on a day of such a small happiness that she might tell him, that she would even understand it to tell. In some way she wishes she could tell him; but also it is good to have the bowl inside her and him beside her and him not knowing.

He does not always stay the night, but tonight he asks if he can—perhaps he sees the spark of the small happiness in Sylvia. Sylvia has never understood why he wants to see her at all. It is he who has pursued her, entirely. Sylvia has neither resisted nor particularly encouraged.

As he stands to undress he is directly beneath the dangling light bulb. He reaches his hands up flapping.

—Can't you get rid of these fuckin' *moths,* Sylvia? Christ.

His skin is utterly pale, his ribs stick out, his back is pimpled. Sylvia has been lying in long underwear and as she pulls it off it runs prickly static along her leg hairs. She looks at her own thin pale limbs. She looks at him. She thinks ahead to the act of sex, which did not always work for them. Plaguey, sterile love. From the calm distance she still maintains she sees that this will be the last time. She will tell him to leave the light on, she will insist that he see her, not as anonymous female flesh, but as Sylvia. She will force his classifying mind to face uniqueness. This is Sylvia whose flesh you touch, Sylvia's legs you move between, Sylvia you try to possess. But you won't. The radiators shudder, wind bangs the loose window panes, the black cloth luffs. From below comes the irregular blurt of TV voices in the silence of Sylvia's lover undressing. He stands with his hand at the waist of white skivvies and reaches for the light cord, and Sylvia says,

—No.

Part Three

The Calculus of Imaginaries

When is identity not identity? Consider, for example, the calculus of imaginaries. The star of the show is *i,* the square root of negative-one. Yet negative-one has no square root, no negative number has. Any two identical numbers, negative or positive, multiplied together result in a positive number. The square root of one is one or negative-one. The square root of negative-one is a void: they call it *i.*

Sylvia explained it as we sat on her fire escape on the evening of the summer solstice. The longest day, and therefore the briefest night. Was that a safe assumption? The trouble starts, Sylvia said, when you assume.

As for that unassuming little non-quantity, *i,* Sylvia said it was expedient in certain equations of a highly theoretical sort.

It was thrown into the trenches and eliminated before the battle was over—you couldn't have so unseemly a figure popping up in the product. Even the calculus of imaginaries, it seemed, finally settled in the realm of real numbers.

I asked the stupid question (there are no stupid questions, only stupid people who ask them): What's the use of theoretical equations, numbers that can't exist?

—What's the *use* of a poem? Sylvia rejoined.

As if I had any special fondness for poetry. Sylvia herself had suggested that I had become a reporter precisely because my father had been a poet. Language was my metier, but as far removed from poetry as could be. Even math apparently was closer.

Anyway, Sylvia insisted that *i* and all so-called imaginary numbers did exist, as firmly as any numbers, which were all abstractions, all imaginary in some way.

—When we say *one* we don't mean one apple, one stone, one book. One is a hypothesis, form without content. A house is air within walls; the world is a structure of ideas around the bewildering flux.

Dramatic voice, rising to a quaver of excitement. It still charmed me that she could be giddied and thrilled by ideas, by her own words that walled them in.

—The patterns we imagine on randomness, I said.

—Maybe, said Sylvia.

Maybe. The point that had fixed the line from me to Sylvia was that Silvia with an *i*. The imaginary number, or the central pronoun, modestly but emphatically, traded for the *y*, the prime interrogative. But that Silvia came from poetry, and what use was that? Well, it had brought me to Sylvia. Her poem, and the Bard's, had conspired in that. *What light is light? What joy is joy?* These questions of identity were perennial.

Sylvia had stopped writing poetry. The need dried up about the time we met, she said. Poetry had been her way of talking to herself, when there was no one else to talk to. Now there was me, there was Carla. It was remarkable for her to spend so much time with people now. She had wondered if she would ever regain the knack of human discourse, and she was only just beginning to trust that it wouldn't be taken away again. It was a great gift, she said, to talk as we were talking this night, not to have to worry her words or her thoughts, as she had when each word or thought had borne the weight that it might be her last.

—Still, now I'm always prepared to be alone, she said.

It was a simple truth, but I couldn't help taking it as a warning. It was her past that made it so, it was her past that made her, whom I loved. It was her past that was my fiercest rival. She talked of her past more gently now, as if she were able to draw her memories, even the painful ones, closer to her:

How her father had come drunk and raging to the house one day after the divorce, and Irene had tried to call the police, but Richard stopped her and calmed Sylvia's father down, and even drove him home.

How around this time she had made two earnest juvenile attempts at suicide: the first time she hid in the garage's open rafters and sawed at her wrist with a figure skate. Never a quitter, Sylvia had managed deep abrasions with the skate blade's jagged toe, but nothing like the garish crimson flow she had imagined. When Sylvia started crying, even as she gouged onward, her mother had come out into the garage. Sylvia, surprised, dropped the skate, and the blade broke on the concrete floor. Her mother said to get down from there right now; that skate would come out of her allowance. (She held out her arm; she still had the scars. Numerous chicken-scratch white lines on her wrist. If they'd been found inscribed on an ancient stone, they might have been called a language; some cryptologist might have discerned a pattern.)

The second time, home sick from school in early spring, she was alone in her upstairs bedroom while her mother sat reading in the living room. Her mother had forbidden television: if you're too sick to go to school you're too sick to watch cartoons. In the silence Sylvia listened to the hum of the refrigerator, the sigh of the heat vents, a stylish new digital clock haltingly flipping its little numbered placards, her mother flipping pages more crisply, briskly—savagely, it seemed to Sylvia—tearing out a coupon or recipe. Occasionally her mother laughed a little, and Sylvia thought she sounded like an idiot. Then all these little mindless sounds in the stillness and all the wretched unfairness of her life brought Sylvia to the edge of an insane rage. Out of her feverish bed she scrambled, and down the stairs she flew. She flung open the front door which swung within inches of her mother's chair, and she ran down the walk in pajamas, bare feet. The day was drizzly, dark; the gutter was full of gray slush, and she threw herself down there, and she thrashed and screamed. The suburban streets were empty. Streetlights gave off a quaint, feeble glow. Sylvia quickly wore herself out—she really was sick, after all—and then she lay there soaked and freezing and dirty, and she cried. Still her mother did not appear. Sylvia picked herself up and went back into the house. Quietly she closed the door and passed behind her mother's chair, back up the stairs. Her mother did not move or say a word.

Sylvia told these stories with gentle humor, harrowing though they were. These were her memories, and they were her, and she was growing, perhaps, a little fond of herself.

Still she admitted that sometimes the walls of her room

closed in on her, and forced her out at night to walk, long late walks back and forth across the river bridges, as she had done compulsively that first winter at the University.

—I wish you wouldn't do that, I said.

—I can't help it.

—When you need to get out, just call me, I'll come and get you.

—When I feel that way, I can't be with anybody. It's not that I'm lonely, it's just that I'm crazy.

—It's so dangerous, I said. It's stupid.

—I know, I know. I'm stupid, I'm defective.

—I didn't say *you* were—

—You have to appreciate how far I've come. It helps having you here. It helps having you. But it won't all change just like that. It won't all of a sudden be fine. And you can't change me.

—I don't want to.

The topic ran aground on the reef of those suspect assertions: You can't change me; I don't want to. I couldn't vouch for the truth of either one. I remembered what I had seen in her the first time we met, what had continued to fascinate me: she was brilliant, and defective; confident and shy; challenging and demurring. I had wanted to protect her from this brutal, monolithic world. Did that mean I wanted to change her? I thought of the secret desire, and if our ideas were the walls of this house that was the world, then the secret desire was its foundation, and it was dug deep in the ground. If the secret desire said, *Change her,* and she said, *Don't,* which would it be? As I found her she was shaped by all that had come before; and if what came before was my rival for her attention, as it seemed, then wouldn't I want to change that shape, just a little? Then I set the discussion on a more philosophical tack:

—Don't you think we alter the world to fit our conception of it? I said.

—Alter is what you do to promiscuous pets. It involves surgery. It's not what you do to people you love.

She faced me with her forthright green gaze, and I could not but concede. Sylvia's house, however abstruse its decor, was built on a footing of cold moral stone. That was one reason that living there was so difficult for her—that and the haunted closets.

We watched the last light fade on the longest day of the year, and decided we would see this brief night through till morning. We spread a blanket on the grating to soften our seat, and held each other as the stars struggled out against the long late twilight. In the middle of the night Sylvia said,

—My family's having a picnic over the Fourth, I mean my whole family, Mom's side, out at Gran's.

—Uh-huh, I said.

—Would you like to come?

—Are you going?

—Of course, if you'll come with me.

—Well, don't go to any trouble on my account.

—I mean, yes, I'm going, and I'd like you to come. You're invited. I'm inviting you.

—Will I have to lift heavy boxes, or break sod, or anything?

—It's purely social. It's a picnic.

Then I said I'd be delighted, though that was stretching it. I had forgotten about my bargain with Gran—I still couldn't quite believe that I owned a '62 Bel Air, though I was credulous enough to drive it. Now remembering it brought a pang of conscience. I wasn't quite sure I had agreed to clean up the whole place, but I had to pay for it somehow, and I hadn't paid at all.

We fudged the dawn, not waiting for the sun but satisfying ourselves with a distinct eastern glow. Then we climbed inside and fell together, to bed, to sleep, all at once, and slept late into the first morning of the new season.

Yardwork

92° the car radio said as I pulled up at Gran's for the Kraus family Fourth of July picnic. The number sparked my interest, more than just to wonder at the heat. 92° was the magic, tragic number, meteorologically speaking. Its portent had its source in some late '50s sci-fi flick: a bleak desert town; danger crouches coiled at the outskirts of the city; insidious infiltrators from outer space, body-snatchers, mind-snatchers. In the suffering heat, in a spare room of sharp angles—why are the windows closed?—the sheriff and the man of science (in a tweed jacket and tie?) and the girlfriend, the ideal '50s woman, sit and wait and speculate. It's the sheriff, an acute observer of human nature, who gazes out the window—there hangs the fervid red bulb with its sky-rocketing vertical—and muses: 92°. *The most dangerous temperature. Cooler than 92, people are easygoin'. Any hotter, it's too darn hot to move, too hot to make trouble. But just 92°, people get irritable, they get crazy. Anybody can crack.* He peers, strong-jawed, out the window at showdown-deserted streets. The man of science and the girlfriend nod with worried looks. They know about 92° too. Everybody does.

That one scene was all I remembered of the movie. And I more than remembered the 92° dictum, I believed it, I had secured it in the cell of my mind that held the essential explanatory facts of life on this planet, and I would remember it, I reckoned, till the day I died, when so much had been forgotten. Already I couldn't remember the name of the girl who had chased and caught and kissed me on the playground in first grade; and the capital of Alabama; and my grandmother's maiden name. I always had trouble with my brother's birthday, and my own phone number, and my mother's favorite color. And I had a hard time believing other facts the factuality of which I would never dispute, which you could look them up: that colds come from germs and not from drafts or wet feet; that certain starlight we see travels from stars that no longer exist; that the earth turns round and the sun stands still; that glass is a liquid. The 92° dictum you couldn't look up anywhere, but it seemed to explain more, more meaningfully. Like so many vulgar errors, the much overrated common sense. But still I pulled up at Gran's (arriving late by myself, having covered the neighborhood celebration for the paper) believing that 92° was the most dangerous temperature, and so I was nettled by a faint unease, and I tried to force a cheerful frame of mind over a vague, swelling foreboding. Not that I ever expected this to be a festive event.

The Kraus cars were parked along the road. I parked there too, and stepped out of the car into the pressing white heat;

the sky hazed and close; the air thick and soaked. As I walked up the drive I tried not to see the overgrown yard, the tall weeds that had gone dry and brown under two weeks of heat. Kaiser lay in the shade of a rusted barbecue kettle tipped against the garage. She was pressed as flat to the ground as she could be; she was trying to be taken for dirt. I came to the edge of the yard unnoticed, or anyway ignored. There were around twenty people in the yard, most of them crowded into a small area near the garage, for the better part of the yard was taken up by a badminton court where several boys in colorful surfer shorts played manly badminton on the sere prickly grass. A boombox on the grass blasted harsh, manic metal. It could have been a beach volleyball game, but behind them stretched the desolate dun slope of tinder weeds. The houses down the slope were now plywood boxes, tarpaper-roofed; I could see heat shimmering up from the black.

Sylvia's immediate family—her mother Irene, stepfather Richard, sisters Rachel and Alison—sat at a picnic table. I had seen a few pictures of them, but I recognized them more from what Sylvia had told me about them, and from the aura of that splintered clan. In lawn chairs and around a round white patio table sat Gran, and the aunts and uncles, and several female cousins roughly Sylvia's age. These girls matched the boys who were playing badminton—cutout dolls, I could see them joined hand to hand and foot to foot. Their hair, chemically blond, it looked like you could just reach out and snap off a handful. Their chubby tanned legs, smooth-shaven, ripply, tapering down to their little fashion shoes, like the shanks of little piggies. Among the older set one man stood out, a thin man with a thin mustache and dark glasses (the kind of glasses that go dark outside and lighten, but never completely, inside). He wore a yellow sport shirt and plaid bermuda shorts, and black shoes and black socks that drooped to reveal his scarred, nearly hairless shins. I pegged him as Irene's chief tormentor, a war criminal, escaped to a private Argentina; he was the well-driller's spiritual twin. I disliked him on sight.

I didn't see Sylvia anywhere, and with a 92° temper I had begun to curse her, thinking—not thinking, merely allowing the notion to arise—that she might have backed out at the last minute without telling me. Then the breezeway door slammed, and I turned to see her skip down the steps. I saw her soon enough to watch her face transform, as I had seen it change when I watched her at Frank's party. By an obvious act of will she replaced her look of weary distress with a bright, open smile. And she came to me and stood tiptoe to kiss me. Her face was flushed but cool and damp, just washed with cold water. Wet strands of hair clung to her forehead.

—Hi, she said. Hot.

—Ninety-two degrees, I said.

—Oh. I thought it was hotter than that. Come and meet the family.

She took my hand in her hot hand and led me into the back yard. The relatives responded so suddenly to our approach, it was clear that, as I suspected, they had noticed me at once and pretended not to. Sylvia introduced me:

—This is Bryce. Gran gave him her car.

Bryce "Gran gave him her car" Fraser. Not man of my dreams, my only true love, fire of my loins. The guy who took the car. And as it turned out, the family was not so indifferent to the fate of that car as Gran had predicted.

—You *gave* the car away, Gran? cried one of the blonds.

—He said he'd clean up the yard, Gran said.

And here came one of the surfer-shorts boys, beer in hand. He peered down the drive at the triumphant neglect and said:

—Nice job.

Then Mr Black Socks rose up, saying,

—What the hell are you doing giving your car away, Mom?

—You were supposed to take me looking for a new one, Gran said sharply. Who wants that car.

To me Gran said:

—Should have done it in the spring, like I told you. Got your work cut out now.

My mind scrambled to concoct a defense, tried to recall that absurd bargaining session. What had I agreed to? Sylvia had been the only witness, and I could expect no help there. There was no question, how thick the blood ran between those two. More important, I knew in my heart that this yard of rank weeds, twisted trees, the dry and dusty tangled bramble, with its hidden insinuations of dereliction and abandonment, this was mine, the cost of my miserable capitulation to expediency.

The matter of the car blew over, no one truly cared. Sylvia brought a beer for me and one for her, and we sat at the picnic table with her family. Irene shared Sylvia's nervously happy manner. Her hair was short and waved, darker than Sylvia's and graying. She wore a blue flowered sundress, and I had to force myself not to stare at her bare shoulders, which reminded me so of Sylvia's. She greeted me maybe a little too warmly, and said she'd heard so much about me. She was charming and a bit flirtatious.

Rachel was stunning and overdressed. Her long brown hair like a shampoo ad. Her features like her mother's and Sylvia's, but with a fragile quality. A flowing print skirt, swirling gray, blue, and yellow, a ruffled white blouse. She was shy. She adjusted her skirt and looked at her lovely slim ankles, her white sandals, flats, her painted pink toenails. Alison was fourteen; I had always pictured her at four. And she was still the baby, though she was already taller than her mother and sisters. She was thin, pale, eager, bright, somewhat frantic. She darted about like a bird.

Richard had big hands. His hair was impeccably gray, his face strong and serious. His clothes were the casual clothes of someone who is never casual: black loafers, blue seersucker trousers belted with black calf and a gleaming silver buckle, a pressed knit sport shirt, white. He wore an expensive watch. He had no place at a picnic.

The rest of the family I met piecemeal. Black Socks was indeed the oldest brother, heir to Kraus infamy, and there was another brother and a sister besides Irene here today. Black Socks was around sixty, and his children were nearly as old as Irene. The boys—but they ranged from genuine boys to middle-aged men, most in their twenties—sold insurance or used cars or were studying business administration. The forty-year-olds seemed twenty-five, and the twenty-five-year-olds seemed forty. Altogether they exuded an atmosphere that was juvenile but not youthful. Is it possible to think of a youthful insurance salesman? They greeted me with exaggerated heartiness, then they left me alone, occasionally throwing a little remark my way, such as, "How about that, Brad." And I said, "Uh-huh, you bet." What on earth were they talking about? They communicated mostly in some secret language of gestures and slang and noises. Much of it seemed to have a crude sexual content. The girl cousins pretended to be disgusted, they scolded the boys. They all referred to each other's boyfriends and girlfriends, and to things that had happened in bars. They had a lot of friends who got in trouble in bars. The parents sat apart. They looked exhausted. Their work was done. They had raised and provided for these lovely kids, and now they were ready to die. Maybe expand the garage, build on a porch, take a trip to Hawaii, as a kind of coda.

A couple of the girls attached themselves to Sylvia, tried to include her in their reindeer games. They all said it had been too long since the family had gotten together, how good it was to see her. They knew all about what she'd been doing, what a genius she had become: how she had been valedictorian of her high school class and won so many prizes; the big scholarship she'd received from that famous school in California; the important research project she was working on—was she going to cure cancer? They had heard it from Irene; she was so proud of Sylvia. They asked about how Sylvia and I had met, and they thought it was a marvelous story, and they all professed a profound love of poetry. Then one of them noticed that Sylvia didn't shave her legs.

—Oh that's so gross! the cousin screeched, but laughing, to show that she didn't think it that gross, but that she was expected to.

—It's not gross, Sylvia said.

—Oh no, it's not, the screecher capitulated. It's natural I guess. But if you had hairy legs like mine, you'd shave them.

—No I wouldn't, Sylvia said. See, I don't shave my pits either.

And she pulled back the sleeve of her T-shirt to demonstrate, and everybody screeched: that truly was gross. Sylvia was loving this, for the moment; she loved being the unshaven weirdo genius whose mother bragged on her. It didn't seem like Sylvia. It didn't sound like her mother.

A couple of these girls were mothers themselves, I gradually gleaned, and occasionally they shouted after their young children, the young cousins who were the only ones who really enjoyed gatherings like this. Two of these kids had fetched a chunk of block ice from a cooler, and they squatted on the brown grass to watch it melt. They had a bet about how long it would take. They kept yelling to ask how long had it been now, what time was it. It looked like it would take a lot longer than either of them had wagered. They expected it to sizzle down like a drop of water on a hot skillet.

Sylvia and the girls ran out of conversation. The boys had collapsed on the grass of the badminton court. The music ended and no one bothered to flip the tape. The dry chirr and thrash of grasshoppers rose from the tall grass behind us, ahead of us, all around us, and suddenly there was not enough joviality and family spirit to dispel the space and heat and drought that encroached from everywhere. The sky was a hot hazy white, dispiriting. It was tinged brown at the horizon by the smoke from dozens of grassfires burning around the suburban fringes, and now and then the sharp scent of smoke rasped in my nostrils and vanished. Gran's land was a perfect candidate. A spark, a carelessly thrown match or cigarette: prairie fire, and all would come to ashes.

—Are we ever going to eat? Gran said.

The women leapt to their feet and dashed into the house.

We stood in line for food. Coldcuts sweated and oozed, mapping greasy gray continents on paper plates. Sylvia leaned toward Richard and said something I couldn't hear. Richard laughed. Sylvia seized the moment and kept insisting on the joke, whatever it was.

—But doesn't it? It looks just like that, doesn't it?

Irene, on the other side of Richard in line, said:

—What? What?

Smiling, eager to be let in. Sylvia said:

—Never mind, Mother. You wouldn't get it. But it's true, Richard, isn't it?

She held his elbow and looked into his face. He walked away with his paper plate, shaking his head. Irene said:

—What? What? *Tell* me.

They didn't. Sylvia left the line too, and that left me standing next to Irene. She put her hand on my waist and said:

—Oh, they're mean. Bryce, what do you see in such a wicked girl?

I smiled, shrugged. I dropped the spoon back into a bowl of eggy yellow potato salad, contemplating putrefaction. When we were seated with our wobbly plates of tepid oily food, Irene said:

—Sylvia, I never see you. Why don't you call?

—I'm hardly ever home. And someone is always using the phone.

—I'm sure they're not *always* using the phone.

—Why don't you call me?

—I do. It's always busy.

—See?

—You could get your own phone.

—Can't afford it.

—I'd like to help out more, but we spent so much on your California fling. All that flying back and forth.

—I flew out. I flew home. That's the minimum possible flying.

—Still, it was expensive.

—I paid for the tickets myself. You gave me five hundred dollars for tuition.

—Money is money.

I calculated: five hundred dollars over four years; one hundred twenty-five dollars a year; about ten dollars a month, thirty-five cents a day—less than bus fare. Money clearly wasn't the heart of the matter.

—If you had just gone to the U, you wouldn't have had to fly. Plus it's much cheaper. God, how much was the tuition out there?

—It costs me more to go to the U.

—It was nearly twenty *thousand* dollars a year.

—Are you listening Mother? It was cheaper than the U. I don't get any aid here. You and Dad make too much money.

—Well, your dad maybe. What's he throwing his money away on now? Sailboats, ski trips, all that macrobionic New Age crap.

—Macrobio*tic,* Sylvia said. What does that have to do with me?

—Oh you've got him wrapped around your little finger, you always have.

She said it with a sort of bitter laugh. Sylvia looked dumbfounded.

—I seem to recall you took a trip to Greece last year, Sylvia said.

—Don't you think I deserve a vacation once in a while? Irene said.

Then she turned to me:

—Where do your parents live, Bryce? I'll bet you call your mother. I wish I had a son.

Richard asked me what kind of work I did.

—I work for a newspaper, I said.

—A member of the fourth estate. What paper, the *Tribune?*

—No, Red Earth *Gazette,* a neighborhood paper.

He nodded, and turned away, and that was the last we spoke. I chatted with Irene. I learned that she worked part-time for a nutrition program that her suburban church had organized. They taught poor people how to cook. They encouraged them to eat beans and rice instead of meat.

—Get them to act like the third world citizens they are, said Sylvia.

—Oh really, Sylvia, Irene said. It's perfectly healthy, and

so much cheaper. It makes them feel self-sufficient to live on what they make.

—It makes them fart, Richard said. The farting poor.

No one responded to that, but the silence wasn't especially awkward, so I guessed that was the kind of thing Richard said from time to time. Beans and rice—I thought of Sylvia's father, the vegetarian. Irene bit into a chicken leg. She licked her fingers and said:

—And I'm a coed, too, just like Sylvia. And Rachel. Did you know that, Bryce?

She was taking extension classes a couple of nights a week, working for a degree in psychology.

—I had to drop out of college when the joy of my life came along, Irene said, turning to Sylvia not with bitterness this time but with a real, warm, loving smile, and I was so confused. —I never got to find out if what everyone said was true, that college is the best four years of your life. Or five or six, in your case, Sylvia.

—It's not, Sylvia said; and under her breath: —It had better not be.

Then Irene turned around and spoke to Rachel, who sat on another bench with Alison, her back to Irene, facing the wide field:

—What are you going to major in, Rachel?

Rachel didn't turn or raise her head.

—I don't know, she said.

—You have to decide pretty soon, don't you, honey? What kind of classes have you been taking?

—Oh, humanities and junk. Is psychology any fun? I might try English. That's pretty easy.

She went to a small Lutheran college not far from the farm where they had stayed during the divorce. An expensive private college. I wondered who was paying for that. She turned around on the bench, and pushing her hair back from her eye, spoke to Sylvia:

—What do you think I should major in, Syl?

—You were always pretty good at math. Why not do engineering?

The very mention of engineering brought terror and dismay to Rachel's face. She wasn't just a pretty, unacademic girl; she was a pretty, unacademic girl whose family expected her to *be* something.

—I only just passed math because you helped me.

—That's not true, Sylvia said.

—It is too. You're good at everything, and I'm no good at anything.

—Do you think you'd like to teach? Sylvia said. You'd be good with kids, wouldn't she, Al?

—Ugh, said Rachel.

—Why not major in hair, Rach? Alison said.

She looked at her sister with perfect innocence, but her bouncing knees betrayed her.

—Shut up, Alison, Rachel said. Just shut up.

The music started up again, heavy metal that sounded hollow, clanking as if it fought against the pressing air. The two cousins with the ice, a boy and girl, had given up their vigil but kept an eye on the ice, came racing from across the yard in panic when they'd forgotten it for five minutes. And then ten, and fifteen. The ice drew my attention, too. It was like a billboard, a movie screen—I couldn't help looking. That small clear peak of ice, jeweled with white stars, dripping cold cold water from its smooth slick scarps onto the hot brown grass—it was an intolerable enticement. Sylvia kept a hand on my neck, or on my knee, or held my hand. Gran warned Sylvia that she'd give me some kind of rash, hanging all over me in this heat and humidity. The humidity was hands-down the most popular topic among the older set, and all agreed that it was this, rather than the heat, *per se,* which made it so uncomfortable. From the other side I suffered Irene's attentions. But not entirely suffered. She seemed to have taken to me. Enough so that she enlisted me in her effort to reform Sylvia's life.

—I wish she'd get out of that awful laboratory, Irene said to me, though Sylvia sat right beside me.—God knows what those horrible chemicals are doing to her, the radiation. Talk to her, will you?

Were the cousins listening now, as Marie Curie became Sylvia the Lowly Lab Lackey?

More and more often from that side I felt a squeeze on my wrist, a hand on my shoulder. She often asked my opinion, or if I didn't agree with something she'd said. For all the attention I was getting, which was flattering, in a base sort of way, I knew that I didn't really exist in this equation; I was just an intermediary, a catalyst.

—Why aren't you going to apply to medical school, Sylvia? Irene asked.

—Who said I wasn't?

—When you become a rich doctor you can take care of me.

—I've always taken care of you Mother, Sylvia said quietly.

A badminton game had just ended. Two boys lingered on one side of the net, awaiting the next challengers. They laughed and scoffed when Alison appeared before them, ready to play. They refused to serve. Alison insisted, and they said, Well, get yourself a partner. Alison called me over; reluctantly I joined her. Alison was defiant; she would not be taken lightly. It seemed to me that the whole family had a huge genetic chip on its shoulder. The boys played lackadaisically at first, clowning as they returned our shots. It took Alison and me a few minutes to get into the game, but the boys had made a classic error of underestimation. Soon we were returning their shots routinely. And now they had a hard time adjusting to serious play. Alison was quick, and when she returned a hard shot they were goaded to hit the return harder, often into the net or out of bounds. The boys started arguing with each other—it was sort of like watching a schizophrenic in self-conflict, maybe. They won the first game, narrowly, while we were still warming up, then we trounced them in the second game, but they came back to beat us in the third. I said that was enough. The boys made an obnoxious show of triumph, but everyone knew they'd barely escaped. They had almost lost to a girl and weird Sylvia's egghead boyfriend. Alison wanted to keep playing. She was angry. She thought we should have won. She played some more after that, but the boys kept bumping her aside, taking her shots away. She was just a skinny girl and couldn't be any good.

Rachel had been sitting stone still all afternoon, as if, what?—as if movement would attract attention and danger, abuse, accusation? Or simply not to muss her dress. Alison, at the other end of the bench, could not suppress the random motion of her thin white limbs. Not that she tried. Her bony knees jittered ceaselessly, her hands drummed the picnic bench, twined in her hair, picked at the dry grass. She proffered inanities to Rachel, so that she might be rebuked for her childishness. She exaggerated her babyishness—an old trick for attention she would use while it worked, but maybe it always would. Though it would not always be becoming. It scarcely was now. Alison couldn't get enough attention, and Rachel wanted none. Alison was somehow left out of the whole family drama. The others left her out intentionally, I think, to spare her. They didn't see how much she wanted to be a part of it. They didn't see what it would cost her to be excluded, like the only survivor of a battle who suffers a life of guilt and the burden of many deaths. Who drew the lines here? Who said who went to fight? I think it was Sylvia. She protected Alison, or thought she did. I sat on the bench and I felt dizzy, from the heat and the beer and the heavy food. And from something I couldn't quite define, and which I thought might be indefiniteness itself—something that arose from the endless, pointless back-and-forth between Sylvia and Irene. We assume that people act reasonably, rationally; we expect them to make sense; I don't know why. They knew what they were doing, I think, and then, did not. Which is to say, they knew what they were doing, and did it intentionally, yet could not do anything else. Oh what do I mean? The manipulation in this family was so obvious, so blunt, so deeply unsubtle—why did it work?

—When I was Sylvia's age I had two kids already, Irene said. I had to grow up in a hurry. You kids are lucky.

Sylvia left my side and went to sit with her sisters, turning her back to me as she went. The music stopped as the tape ran out, and Gran decreed silence. The boys groaned but obeyed. Silence she decreed and she received, the dry dusty silence of the fields. Haze thickened in the sky, so it was almost clouds. The ominous minor-key skirling of the insects in the field returned, and rose to a maddening pitch. The air was oppressive; it smelled of dry weeds and dust, and the stale clash of perfumes that the girls emitted. It made me think of the cedar closet, that thick scented air. Welcome to the torturers' picnic. Choose your weapon.

Why was Sylvia here? Why was I? Why was I surprised to

see Sylvia and her mother set upon each other, as if by reflex, as if the timekeeper had rung the bell and the referee called, *Fight!* Maybe I had thought that Sylvia exaggerated their difficulties, that she and her mother shared the ordinary, flawed parent-child bond, which Sylvia's perfectionism could not abide. I was wrong. I tried to puzzle it out. I pushed at the recalcitrant pieces, turned them over, trying to force a fit: the divorce, a bad and messy one—traumatic, yes, but nowadays all too common. The cedar closet, but that was more symbolic than anything; Sylvia had been shut in there only rarely, through ordinary childish cruelty. And she had admitted to me that had it not been for her mother's precedent, she would probably have forgotten all about it. The abandonment on the farm—but hadn't that been one of Sylvia's happiest times? But of course it was deeper than any single event, any string of events. In my mind the puzzle into which I was forcing misshapen pieces was a picture of Sylvia's family. The picture on Gran's dresser, the one that had struck me on my first visit here. The odd disconnected figures, touching each other as if the contact were distasteful, or as if the contact wasn't real, a photographic trick. That picture: Sylvia standing where her mother should be. Then the pieces fit. The tone of their talk, something in it had bothered me. They conversed as parent and child, all right, but the roles were reversed, and the child was a three-year-old. There was no right answer to the questions, no response that would halt the inquisition. But of course it wasn't answers that were asked for. The child crying *Why? Why? Why?* and *Can I? Can I? Can I?* isn't looking for an answer but only for attention, to not be ignored, forsaken. *I've always taken care of you Mother.* Sylvia blamed for the sins of her father. The family cycle goes on: Sylvia had herself a child unexpected, before her time; just like her mother. Irene always the victim, so how could you blame her? Irene throwing the lid off the boiling pot, letting the guilt fly everywhere, because she couldn't be touched.

I looked across at Sylvia and caught her eye. I stared at her. What I felt was awe and pain, pity and love, but her expression as I continued staring, speechless, said that what my face showed was the most abject stupidity. Or she saw all I felt, but rejected my understanding as too slow, too late, too little. Or did she think I had sided against her? I hadn't. I had tried to be pleasant and uninvolved. It was a *picnic,* damn it. But Sylvia was angry with me. I would accept no blame; I knew I had been loyal. Was I to blame for not having heard, at first sounding, the silent mayhem over debts of loyalty, love, and sacrifice?

I turned away. Irene sat on one side of me. The seat on the other side, where Sylvia had sat, was empty, but I still felt pulled from both sides. It was then that I left the picnic and went looking for dangerous hardware. Someone had to pay for something.

How satisfying it was to swing the rusted scythe, feel hanks of weeds tear away in my hands. In the garage I had found the scythe, an axe, trash bags. *Thwack!* Cleanly severed, cousin Carl's head bumps and bounces onto the coldcuts tray, goes plop in the Miracle Whip. Whack whack, and down go the Johnson girls, I've cornered them by the rose trellis. Chop chop, bloody red, bloody rose. Red blood, white rose. I show no mercy; neither do I kill with rancor. What's just is just. Black Socks lies face down in the weeds. He must suffer, for his own sins, for Grandpa Bastard's, too. Force his face into the dirt. He should be staked to perish torturously, bitten all over by chiggers and ants. And then the Stenmarks and stepfather Richard. This part with ritual solemnity. Offer them *hara-kiri* with the plastic picnic utensils. Though I would like to do Richard myself. Call it regicide, smug bastard. And Sylvia, darling, you'll have to go, too. You last. You understand.

What about Gran? I'll just clap my hands, say: *Boo!*

Then free the dog, and we lope away across the hills, leaping high, crying out: *Aah-ooo!* Beyond the hills, another subdivision. Shit.

It's a scientific fact that more murders are committed at 92° Fahrenheit than at any other temperature.

Alison came down the drive dangling a badminton racquet loosely in her thin fingers. Her face, of skin so pale that blue etched her temples and cheeks, was mottled pink. Could I spare just one? I should like to see what this one becomes.

—What a bunch of assholes, she said twirling the racquet.

With hands on hips she considered my progress and said,

—You don't have to do this, you know.

And I thought: Oh yes, I do.

—Might as well, I said.

—It's too hot.

—Be hot all summer.

—Not this hot.

I looked at my watch. Fifteen minutes had passed since I started. I would have guessed an hour. Already I was dripping with sweat. Behind me on the drive there was a sizable pile of weeds, and another of trash I'd found hidden in the yard. Ahead of me there was a small patch cleared of weeds but not much the better for it.

—This is gonna take for*ever,* Alison said.

I wiped my grimy forearm across my dripping brow.

—The longest journey begins with . . . , I said, and left the cliché hanging.

—Want me to help you?
—You don't have to.
—I will.

I set Alison to gathering trash from the weeds. I hacked onward. The scythe was so rusty and blunt that I was forced to grab a handful of weeds, pull them taut, hack at the stems until they gave way, several swipes for each little clump of weeds. Bending at the waist this way, sharp pains shot up my back. I had wandered into a patch of stinging nettle, and my bare legs were burning, inflamed. My T-shirt was drenched so I pulled it off and tossed it aside, and it seemed to me that my pale skin began to burn instantaneously. Would I give it up to return to the picnic? I did not even consider it.

Alison was digging out amazing rubbish. Along with the expected cans, bottles, fastfood packaging and cigarette packs, her searching brought forth a rusted bed pan, a golf ball, a baseball with the cover half off; an eight-track tape—Deep Purple; a junior high biology textbook with the front cover ripped off and a copy of *King Lear* stamped all over as property of the local school district; several mail-order catalogs (one offering sexy lingerie and marital aids); a tailpipe, a hubcap, a whole stack of Jehovah's Witnesses literature (as if someone had lost the faith on Gran's very doorstep, chucked it in the weeds); a stocking cap in the University's colors, one knit mitten, one sock; and a headless, armless, one-legged baby doll, the kind that wets. Alison screamed, or gave a kind of yelping whoop, and came leaping toward me, her hands held up and clenched. She was shuddering, shaking her head.

—Oh, gross gross gross, she said.

She pointed. I walked to where she pointed. It was a hollowed spot in the weeds, as if deer had slept there. But deer don't drink peppermint schnapps. The empty bottle lay there, along with a quart malt liquor bottle, and right in the middle, two sticks stuck into the ground, over which two condoms were draped like some kind of weird mushrooms. Flags on the mountaintop. It also appeared that someone had thrown up there.

—Fun time Saturday night, I said.

—Right outside Gran's window, Alison said. Yuck .

—That's young love, I said, and Alison struck me.

—That's not love.

—Well, how would you know?

I uprooted the little rubber trees by the trunks, and was going to shove them down into the schnapps bottle, but Alison screamed again, and hit me again. It wasn't hard to find a paper bag, and I deposited condoms, sticks, and bottles in there. I tossed the bag on the pile, and went back to my cutting.

—If that isn't love, what is? I sang, missing the tune entirely; Alison gave a dying moan.

Now as I worked Alison shadowed me, circled me. Couldn't she ever sit still? Her recent find had put her off trash gathering.

—How do you like my family? she said.

—Oh, well, they're, you know, interesting.

—Totally, totally fucked up. How about Richard?

—He's okay, I guess. A little stiff.

—A total prick.

—Such language.

—Puh-lease.

—Okay, he's a prick. A little stiff prick.

—Oh, puh-*lease!*

—You started it.

Chop. Richard is the target; the specifics of the assault are unmentionable.

—Oh god, Alison said, and she started walking in circles, sort of hopping, her hands clasped in front of her.

—How much do you remember, from when you were little, the divorce, and staying on the farm? I said.

—You know all about that.

—Sylvia told me.

—Of course she did. I don't remember much. But they all talked about it endlessly in therapy. They're such hypocrites.

She said *hypocrites* as if she'd just learned the word.

—What do you mean?

—It's like they don't want it to change. It's like, "Okay, now we understand it, so it's all better now." But nothing ever changes. They like it this way.

—Who does?

—Sylvia and Mom.

—What about Rachel?

—Rachel doesn't have a clue. She just wants to be popular.

—What about your dad?

—He's a space case. Mr I'm-an-alcoholic, Mr Nuts-and-berries. I like Dad.

—You think your mom and Sylvia like to fight?

—Their purpose in life.

—Interesting.

—Sickening. Why don't they just *stop?*

She threw her arms down at her sides as she called this question to the hot white sky. I stood up straight. I blinked hard, and dropped the scythe. She didn't mean the question. The self-conscious drama in her gesture floored me.

—Why don't they? I said.

—Why *don't* they? she said.

—Oh why oh why, I said.

—Oh why oh why oh why, said she.

She picked up the golf ball and flung it high in the air, her lanky limbs all flailing with the effort. We watched the ball arc over the yard, and over the cars parked along the edge of the yard, and drop to the road. It hit and bounced high, and dinged off the fender of a big jeepish wagon, then skittered in a crazy hopping dance over the hot pavement, going *pink pink pink,* and rolled to rest on the opposite shoulder.

—Oh why oh, I said.

I took the tattered baseball and threw it with all my strength, and it flew with the torn cover flapping like a bird hit by a cannonball.

—Oh why why why oh, said Alison.

She sat down in the weeds. I reached for the scythe and my right shoulder hurt like hell.

Three of the boys came down the drive. Having drunk copious amounts of beer and exerted themselves all afternoon in the stifling heat they were frightfully red. They looked as if they might explode. I imagined great splattering aneurysms.

—Yo! Buh-ruce! one of them said.

—Hubba hubba, guy. What a bod.

—Lookin' good, bud. Have this lookin' tidy by winter.

They gave me advice on fixing up the car. Big tires, mag wheels, hood scoop, metallic paint job, dual exhausts. Then one said:

—Nah. It's perfect for Bruce just the way it is.

They laughed. I had to laugh, too, because the scythe was itchy in my palm.

—Why don't you guys step right up here and have a whack? I said. It's a lot of fun. Dollar a pound.

—Dollar a pound. You pay *us*.

I was tempted. After all, the car had been free. But they lost interest and wandered away. I looked for Alison and couldn't see her. Then her head appeared above the weeds.

—Are they gone? she said.

Alison set to separating the bottles by color and lining them up on the drive in order of height. There were a couple of old-timey Bubble-Up and Orange Crush bottles that to her were truly like archaeological finds. When Sylvia had said the yard hadn't been mowed in years, I had thought she exaggerated about that, too. From now on I would know better.

I was itching all over, covered in sweat and dust and bits of leaf and seed. Sweat poured off my brow and stung my eyes. I took a break. I found a faucet hidden by vines next to the front steps. The water gushed rusty, then cleared, and I scooped up handfuls and threw water all over me, an unparalleled delight. I ran my red nettled legs under the cold water. I felt a bit like William Holden in *Picnic,* working bare-chested in the yard, with the sensitive younger sister allied with me, an outsider, the underdog. But if this were *Picnic* and I were William Holden, it would be Labor Day and not nearly so hot, and I'd be wearing dungarees and boots so my legs wouldn't be on fire, and my torso would be tanned and muscular, and I would have a natural, if somewhat crude, way with people. And, I would be in love with Rachel, not Sylvia. Sylvia wouldn't exist.

I wasn't William Holden, and it was Rachel who scarcely existed. She sat there looking at her ankles and touching her hair, and adjusting the ruffles at her wrists. She sat there picturing herself as a confident, composed young woman. She sat there trying to convince herself that if her exterior was just right, she would be beyond criticism or hurt, unassailable.

Inside, okay, she's a mess, but what are curling irons, make-up, fashion for? If the surface is impeccable, perfect, then only she will know all that's wrong within. But surprise! Only she cares! Truly, she hated herself. And she was so muddled that she hadn't even gotten around to blaming her parents for it. Such arrogant omniscience is his who cuts weeds. I thought maybe she blamed Sylvia for it, a little, because she clearly admired Sylvia, wanted her approval, wanted in some way to be like her, but there was no chance of that. Irene doted on her—did it take a child so helpless to make her mother? Richard seemed to look at her as a lost cause. Weak, silly girl. Could someone love her? Sure they could. William Holden would love her.

I was standing under the front picture window, with the brown and yellow curtains. Inside, in the dusky silence, Hermes feigned take-off, and Jesus hung on the wall. Upstairs, down the white hall, the cedar closet. I thought of Grandpa Bastard, way up in the sky, looking down on this misbegotten gathering. He smiles.

I couldn't bring myself to go back to cutting weeds. But I couldn't leave the yard like this. The apple trees drew me as some kind of salvation. I picked up the axe and started down the drive.

—What are you doing? Alison said, looking up from her sorting.

—Like Rimbaud, I intend to fell the tree of good and evil, I said, I think.

Alison looked at me funny. I was pretty sure it was Rimbaud.

I circled the four dead trees. Squirrels, squirrels. What could squirrels have to do with it? I could see into the back yard. I saw Sylvia and Irene sitting at opposite ends of a picnic bench. They both sat with arms and legs crossed, as if they were cold. They were turned slightly away from each other. The funny thing was, they seemed to be talking to each other. The cousins communed through their strange, belligerent incantations. The music had returned. It battered on vacantly. Richard approached Irene. He thrust out his wrist to display his expensive watch. Irene smiled and touched his arm. He gestured impatiently and turned away.

Why didn't they just *stop?* In college I had read a book by an anthropologist who had lived among a certain jungle tribe, and in spite of the generally dry, scientific tone of the book, one gradually came to realize that the author detested the people he had studied. In little lapses from the high tone of scholarly observation it came through that he found them dirty, cruel, indolent, crude—*shiftless,* if anyone ever was. He found no evidence of love, compassion, affection. The men abused their wives, mothers neglected children, children tormented one another, brother betrayed brother and father conspired against son. They lived in lowest poverty, barely subsisting in an abundant land. There was no clear reason why they remained together, but the tribe had existed since the Stone Age (which, really, they had never left). They were a tribe, not shiftless individuals in the rain forest. Why? Why? the anthropologist asked. Why didn't they just *stop?* And all he could come up with was that any tribe is better than no tribe, and that you remain in the tribe simply because it is your tribe, because someone alone is no one. Your tribe is your context, without which there is no reality. To leave the tribe you must face the challenge of a new reality, or the possibility that nothing again will be real or meaningful. That's why you don't just stop. The tribe provides this most basic existential protection, and extracts its payment in blood.

There sat Sylvia and her mother, Irene and her daughter. Without each other, who would they have to blame? Maybe half the crazy people living in the streets had found that fate by saying *Stop. Enough.* And without the identifiable enemy there was a conspiracy down every alley. Irene had suffered her share, and she wasn't about to suffer it silently, or alone. And whom could she blame for opportunities lost, for motherhood at nineteen, for a nightmare marriage? Maybe she really had wanted to be a nun, and had just been trying on sin to see what she would be renouncing. But sin sticks. Or just call it life; it drags you in.

They say the kidnapped grow dependent on and protective of their captors; torture victims feel a certain tenderness for their tormentors. These are strong human bonds, after all, maybe stronger than any they have known. Bound this way, can you not see that the ties are deviant and destructive? Or do you just not want to?

I swung back the axe and drove it toward the trunk of the first tree. I hit the trunk too high, and the frail tree shuddered and whipped and threw the axe from my hands, and I leapt away as the axe flailed the air like a loose firehose and then dropped in the weeds. My hair stood on end; I quaked and shuddered too.

—Jesus fucking, holy shit, I muttered as I went to retrieve the axe.

—What happened? Alison called.

—Never mind, I said. Nothing.

My hands were stinging, but I shed the cotton gloves I'd been wearing for a more intimate grip. Chop down, and low, I told myself. Woodsman, spare that tree. Not a chance. I set myself, drew back, and swung, and this time the blade drove deep, and I pulled it back with a satisfied sigh, practically a groan of pleasure. Three more swings, and this tree was cut half through. There were three trees after this; I should savor every blow.

The two young cousins with the bet about the ice had come partway down the drive to watch me cut the trees. They hung back as if they could read my thoughts, could see what I imagined as the blade found yielding tree flesh—they say children are especially sensitive. And the boys were watching too, while pretending not to watch. Now they wanted in on this, but they'd missed their chance. Two trees were down, and stripped of limbs. I would need a saw to cut up the longer lengths.

Alison had taken up the scythe, briefly. Her progress in the yard was negligible. Now she walked over to me examining her hand, saying:

—I think I'm getting blisters.

Not as if she cared, merely curiously. I let down the axe and leaned on it, a regular Paul Bunyan. Alison and I looked toward the back yard.

—I hope we can go soon, Alison said.

—Aren't you having fun?

—Tons.

The heat and, perhaps especially, the humidity were wearing on the crowd in the back yard. Those who moved moved slowly, with effort. Those who sat appeared to be melting. Clothing wilted and sagged, revealing the real imperfect shape of everyone. Even from here I could see, with some pleasure, how Richard's white shirt stuck to his chest, showing flabby breasts, and clung at the handles of his waist. Sylvia and Irene were bookends on the picnic bench, but one was youth, and one was middle age and motherhood: Sylvia's clinging blue T-shirt was enough to arouse me at twenty paces, the way it was damply drawn to her breasts and flat stomach; Irene's sundress showed her low motherly bustline and bowled her belly, but she leaned back unselfconsciously. There was a kind of bravado about Irene that I admired. She had to know that she could be unlikable, but she made no compromise. There was something of that in Sylvia. Only Rachel seemed unaffected by the oppressive atmosphere. Her blouse stayed crisp, her hair buoyant, the folds of her skirt miraculously full. The icy will that Sylvia and Irene turned to each other, Rachel focused on her appearance. With Alison it didn't matter. In her powder-blue sleeveless blouse and black gym shorts, she was just all there, always. And there wasn't anything to be revealed in her girl's stick figure, anyway. And me, I was a mess. My shorts slid down my hips and chafed, my socks slumped over my ankles. Except for Rachel, all pretense had dissolved in the muggy air.

Then as I stared, lightheaded and exhausted, I saw everyone slumped naked, or wrapped in rags like refugees. It was not a picnic but some terrible ordeal, a placid world turned upside-down by fate or history. Under this inhuman hardship, humanity fled; cannibalism would soon commence. I looked away. Alison reached over and touched my neck. I cringed.

—Oh god you're sunburned.

I knew that quite well. I dropped the axe and went to find my shirt, and as I pulled it on, I thought: Then what if, instead of mass murder, I go bursting into the picnic (I'm a great brawny giant of a man, my skin glistening and taut), knocking cousins sprawling with casual swipes of my forearms (forearms like Popeye, but biceps, too), throwing aside picnic benches and lawn chairs (Jesus rousting the money-changers came to mind, but didn't seem appropriate), and gather up Sylvia roughly in my manly arms ("How dare you manhandle me!" protested the comic-book beauty), and stride away with her, and lay her down, and take her in the weeds; she is ravished, ecstatic, now she knows what truly matters.

What if?

I looked at the front yard, the drive, and what a god-awful mess it was. It was cut halfway across, and unevenly, there was a fringe at the edge of the house where the weed trees and vines had found firmest root. At least before there had been an inconspicuous consistency to it. Just as in the cedar closet, I had stumbled in with no idea what I was getting into, and I'd fucked it all up. My heart sank, I despaired. I dropped the axe and moved out into the drive, hanging my head.

—*God!* Will you both just fucking *stop?*

Rachel stood by the picnic table. Sylvia and Irene faced her. Her hair fell around her face, she threw her clenched fists down at her sides.

—*Can't you both just fucking stop it?*

She ran toward the house while Irene said something about your language, young lady. Halfway across the yard Rachel slipped and stumbled, made the breezeway steps, ran into the house, slamming doors.

No one moved or said a word. Even the insects in the field seemed to have ceased their keening.

The two young cousins who'd been watching me suddenly looked at each other and raced into the back yard, stopping just about where Rachel had stumbled. They twirled about, frantically searching the grass. They dropped to their knees and flung their four palms to the ground as if they'd choreographed it. Then they held up their hands, which I guessed were wet and a little cool, and they wailed. The ice was gone, but when? They yelled what time is it, what time is it? How long did it take? But of course no one knew, and both insisted they'd won the bet, and they were nearly hysterical with disappointment and frustration. Now Sylvia came down the drive, walking purposefully, not as if to visit. She hadn't come to see me or shown the slightest interest in this yard-clearing the whole time I'd been out here, which was how long? Only about an hour and a half. I might have pretended offense, but the time for pretense was past. She didn't stop walking and only turned her head briefly, not really looking at me when she said:

—I'm going home.

I nodded. I said:

—Okay.

There was much understood here but little to be done. I knew she was angry with me and that she knew there was no cause. She knew I knew that she felt guilty about how she was treating me, and for having brought me into this, which was why she couldn't look at me. Transference of aggression—it was a term Sylvia had used once, it referred to something animals did, and I was privileged to understand that it was working here in this world of animus and unreason. And I knew that even though she and her mother probably went around like this every time they met, she was far from used to it, and the strife had opened wounds. I don't know if she knew I knew that. And I must say that though I knew this, at the moment I didn't really care, because I was pissed with Sylvia, too. I'd been brought to the battle and used as a tactic, expendable troops, mother fodder. And maybe she saw me as a turncoat, but mutiny is the due of cynical generals.

—I'll come to your place later, I said.

She shrugged. With her back to me she said,

—Whatever.

I heard her voice quaver. She got in her car and drove off.

The young cousins now came to blows over the ice. I heard a shout and turned to see them tumble in a mutual headlock to the grass. Black Socks rose from his chair and moved toward them, yelling,

—Goddammit can't you behave for once?

And he grabbed the boy's arm and literally threw him away from the girl. He hadn't even set down his beer. He returned to his chair and sat down. The older girls then took the girl aside, and the boys went to the boy, feigning punches, mocking him, saying what are you doing fighting with a girl, pick on someone your own size, though the girl in fact was bigger. The little ones were still wailing, and I could see the tracks of tears down their red grimy cheeks. Of course it was no one's fault; still this terrible disappointment had occurred, and who else could they take it out on. They weren't really mad at each other.

Now the picnic started to break up, and groups of the relatives came down the drive. I moved to the far side of the yard to avoid them. The Stenmarks were among the last to leave. Irene came down with her arm around Rachel. Rachel's head was bowed, her face hidden. Irene stopped to say how nice it was to meet me. Richard passed, saying:

—Bucolic Bryce.

Last chance, I thought, but I let rest the axe. Alison paused to pet the dog, then came lagging behind the others, and stopped when she reached me. She called to her mother, who was already in the road:

—I'm helping Bryce clean up the yard, can't I stay?

Irene turned, and shielding her eyes called back:

—How will you get home?

—Bryce will take me.

—You've been in the sun too long. I think you'd better come with us.

—Oh please, Mom, I'll put on a hat.

—Bryce doesn't want to have to drive you all the way home.

—He doesn't mind, Alison said, looking desperately to me.

I shrugged, and turned half away.

—Come on, honey, Irene said.

—Oh, said Alison, stamping her foot.

—Thanks anyway, I said. And thanks for the help.

—See you, Alison said, and skipped down the drive as Richard started the car and backed into the drive to turn around.

Then I was alone in the yard. The haze had thickened to a gray mat, blocking most of the sun. I walked up the drive and into the back yard, and found the garden hose where it was dribbling into the rose bed. I turned up the water and drenched my head, my face, slicked back my hair, drank long from the hose, feeling as if I would never want to stop. Then I rinsed my arms and legs, turned the water down and replaced the hose in the rose bed. Gran came out of the breezeway door as I walked back around the house. She handed me a sandwich, ham on white bread, with bright yellow mustard and mayonnaise oozing out the sides, held in a blue-flowered paper towel. I accepted the sandwich and took one bite, and finished it in two more.

—How about a can of pop? Gran said.

—Yes please, I said.

She went back inside and returned with a can of Seven-Up and one of her daughter-in-law's brownies. She said:

—Ninety-six degrees, the radio just said.

I nodded, but I knew better. We both went down the drive and looked at the yard.

—It's still pretty much of a mess, I said.

—Yes it is, said Gran.

—But, it's getting there.

—Getting there, Gran said. Ways to go yet.

—The apple trees are down, I said, indicating the brush pile across the drive, far more unsightly than trees in any state.

—Be some work taking out the stumps, Gran said.

—Oh, the stumps, I said. I guess I'll leave that for last.

—It'll be a bear. Stumps always are.

—I've never taken out stumps before, I said.

Gran didn't seem surprised. I said:

—I'm pretty worn out, but I'll work a bit longer. It should go better when it starts to cool off.

But I knew that wouldn't be for hours.

—Going upstairs to lie down, said Gran. There's coldcuts in the fridge and bread in the breadbox. Help yourself.

I went back to work in front of the house. Without distractions from the back yard, or conversation with Alison, I entered the rhythm of the work, bending and swinging in comfortable cadence. My T-shirt, heavy and wet, chafed my burned shoulders, and sweat streamed copiously down my face, but I didn't mind, it was all a part of it. The silence ceased to be ominous and became instead comforting. The whole nightmarish picnic seemed ages ago. I couldn't trust my judgment. Was Sylvia's family really that bad? Maybe Irene was just young at heart, and a little meddlesome. Maybe Sylvia took it all too seriously. But I didn't really want to think about it: It's not your trash, I told myself again; don't go cleaning it up if you're not asked. So I swung into the work, reaping this worthless harvest. And the weeds piled up in a dry stack that I wished had borne some fruit. But that was the definition of a weed, wasn't it?, something that grew without asking and gave no benefit but in fact caused detriment. There were no trash surprises; Alison had done a good job of rooting out the litter.

In a little more than an hour I'd finished clearing the yard in front of the house. While I still had momentum I went across the drive and roughly cleared a patch around the tree stumps where I could pile the brush somewhat neatly. Then I took a break, went in the kitchen and got a can of pop from the fridge, sat on the breezeway step and smoked a cigarette.

It was late afternoon and the hottest part of the day, though the sun was not so strong now as it had been earlier. I knew I wouldn't be able to work into the cooler evening, which anyway wouldn't be much cooler. I had already done too much, and would pay dearly tomorrow, if not sooner. When I'd finished my cigarette I went into the garage and found a wood saw. I sawed the felled trees into lengths I could stack. Now every motion was an effort. I worked slowly, and got the cut wood in some order, then went back across the drive and began taking out the largest of the trees rooted at the house foundation. I had removed three or four, and then I knew I could work no more. I gathered the tools and returned them to the garage, and walked slowly down the drive. The yard didn't exactly look good, but I guess it looked better. There was still a lot of work to do on the yard, but I didn't know if I would do it. It felt like something I might do, and then, might not. It depended on a lot of things. It depended on Sylvia, and me, and maybe Carla, and on other factors I couldn't see to consider. I think it didn't matter that much to Gran. I thought she understood all this much better than anyone, maybe because she didn't try to understand, just knew it, and saw it happen, and because it happened according to some plan, whether or not she understood it, approved. Oh, for the faith in how things happen, and the will to accord oneself to them.

Fireworks

I drove to Sylvia's. I thought we should mend fences now. I thought we had to say something about what had happened today, or things might be changed forever. I might have gone home and sat in my apartment, watched the long gray twilight simmer down, listened in the deep sheets of sound pouring from hundreds of open windows, felt the hot damp stillness drift like clouds of cigarette smoke. And then tomorrow or later, gone to see Sylvia, and resumed whatever it was we had wherever it was we'd left off. But as I say, things would have been different. Whatever you do, things always would have been different. I didn't want to fall into this dumb awe at the way things happen. I wanted to live intuitively, I wanted to find the wisdom in what was. I didn't know how, and so was afraid, and so went at it with bluff assurance, half-assedly, no way to do anything.

I parked on the street and walked through the alley to the back of Sylvia's house. Her car wasn't in the dirt lot, which was now just ruts in the crabgrass—and how could I possibly have felt the urge to fall to my knees and tear at those hard-rooted gray-green nebulae? I resisted.

There was no one around. There were no signs of life at all. Music played from here and there, I thought I picked out Billie Holiday, and a few cars went by on the main street in front of the house, but these sounds were nothing more than background to the utter and universal capitulation to this air that hung like steaming rags.

I climbed the fire escape heavily, one insurmountable step at a time. I did not duck down at the second-floor window, where the northern sun angled in on a worn runner of dusty purple and blue. I didn't care if anyone heard me going up, and I couldn't imagine that anyone who heard me would care, either. Could anyone escape this tropic sloth?

I pushed at the door and it opened. I stepped inside. The room felt so empty, so still. And dizzyingly hot. It baked beneath the rafters. It was prettily lit by the soft sun through the haze. For a few weeks the sun at its northernmost would reach Sylvia's windows. I opened the hallway door, but as there was not a breath of wind, that didn't help. Sylvia had an old oscillating fan, the kind that could chop off a baby's hand, and I set that going, and I guess it helped a little. I wandered around the room, picked a towel off the bed and hung it in the closet. Looked out the doorway at the green treetops, the shimmering rooftops, at the blurry yellow sun halfway between horizon and heaven. The glow softly suffused, spreading bright shadow, illuminating housefronts pale yellow, lambent brick, blazing on a high dormer window. I felt horrible. Why did everything have to be so beautiful? I chanced to look in the mirror over the bureau, and saw that I looked horrible too. The towel on the bed had been maybe a hint. I retrieved it and went down the hall to the bathroom. I had shed my shoes and my bare feet were sticky and made a sticky sound on the wood floor. Rolling heel, sole, ball of foot. The floor felt warm but cooler than the air. The bathroom had no windows, and I couldn't stand the harsh vanity light, so I left the door open. I felt so utterly alone, no one could bother me. The women on Sylvia's floor were aware of me, and didn't care. Mrs Simic, in her perpetual black dress, was probably dead in this heat. I spared a moment's mourning for her as I turned the cold water on full and stripped in the soft wedge of light through the door, and stepped into the shower. I drew a sharp breath as the cold water hit me, shuddered as it raised gooseflesh. I groaned, whimpered. Even under the cold water, I could feel the red heat of my sunburned shoulders. I turned slowly under the water until I was uncomfortably cold, then slowly washed my hair, carefully, thoroughly soaped my body. I wanted to feel the cold creeping into my organs, I wanted to feel ice in my kidneys. I rinsed long and then turned off the water. In a small concession to house rules, in a tribute to the memory of Mrs Simic—God rest her soul, now her plump old body baked like a sausage in that casing of black gabardine—I dried off behind the translucent curtain and not before the open door. But I walked slowly back to the room wearing only the towel, and I felt the air condensing on my cold skin, as if I were a beer bottle right out of the fridge.

I rummaged in Sylvia's closet and bureau, and found a pair gym shorts that fit me. I rummaged in her refrigerator and found a beer. The gym shorts were in the University's colors, maroon and gold. The beer was a long-neck bottle of a cheap local brand. I dug my pack of cigarettes, a rumpled dromedary, from the pocket of my own sweaty shorts, and went out onto the fire escape landing. Since all things are relative, the air outside felt cool. Already sweat welled from every pore, though my skin was still cool to the touch. I sat on the landing in my shorts with my beer, and smoked a cigarette, and felt like any fraternity boy on earth, and I liked it. A pleasant ache murmured in every muscle, and I understood how if you worked with your body every day, all day, then you came home and you just didn't care. You passed the evening somehow, and slept, and went to work again and wore out your body, and went home and just didn't care. Most likely you drank. Your body was tired and wanted it and liked it. It put your mind in the same state as your body, and the point was not to care. Grandpa Bastard's problem was that he cared. He worked the farm and he drank, as was proper, but even then he had too much energy, and he wanted, god knows why, to be a lawyer, and he couldn't, so he got mad. And instead of assaulting the bar association he went after the kids and maybe his wife. If they'd had TV, would everything have been all right? He didn't know what he started. Was he responsible for everything that followed? The connection seemed tenuous, and the feud between Sylvia and her mother appeared quite self-contained. But then nothing is really contained. I found myself thinking that I couldn't imagine either Irene or Sylvia allowing herself to be treated the way the uncles, and then cousins, had treated them; but of course that was all wrong. I mean, it was hard to see how that had happened to two strong women, but they had just been little girls, and they became what they were because of what had happened, everything that had happened, good and bad. Does suffering excuse anything? This is, above all, an inexcusable world, though, perhaps, a defensible one. Irene acted as she did out of what she was and how she saw herself. She couldn't do otherwise. Could I? Do we own will? Yes: we can flex will to become what we are.

I didn't really want to think, and these thoughts were desultory. I did begin to wonder, as the sun was three-quarters to the horizon, and my beer was warm and almost empty, where Sylvia had gone. I couldn't do anything about Sylvia, but I could do something about the beer, which was to open another, which I did.

With a fresh beer in hand I became a little hopeful. I looked forward to seeing Sylvia, to turning the bad day around to good. I thought of the night of the solstice, and I thought that tonight might be like that. Forgive and accept. All that was understood but not said, it need not be said. Because we understood. The darkness of unreason and the fearful distance of souls could be annihilated with a kiss, with the press of flesh. Wasn't that what we hoped? Tonight I chose to believe it.

I woke from a doze on the landing, and started with fright when I saw myself way up high with nothing below. I'd had a dream of glaciers, lordly and blue, and of a fire at an ice house. I had recently interviewed an old man who had once worked harvesting lake ice in the city. They hauled the blocks of ice into wooden barns, and stacked them with sawdust around each block. The wet compressed sawdust could start to give off methane, and if there was enough it could catch fire, and if it caught fire it could burn for days. The old man had seen such a fire. The dry wood barn surrounding the ice had gone like straw. Behind it stood the two-story bricked square of ice, etched every chink and crack with pale blue flame. The memory still held him in awe. He said "dat" for "that" and "ting" for "thing." Working on the ice, he said, the cold crept up through your feet and legs and inhabited every joint. In his quiet room with all the windows down and the curtains drawn, in the still hot air, he sat with his big shoulders hunched and his big hands curled in his lap. He looked at his hands.

—All the fellows got arthritis real bad, he said.

All de fellas got dee art-rite-iss real bad. Obsolete jargon spilled from his loose wet lips like an incantation—*kerf, script, hookeroon, needle bar.* They tended the waters as a farmer his fields.

The air did not cool so much as thicken as darkness slid westward. The smoke from grassfires reddened the sunset. I thought of the two young cousins mad with rage over something that wasn't there. They were just little kids, and yet the rage was real. I was sore and stiff from this tense nap on metal grating, but I found myself too exhausted to move. My left hand was still wrapped around a warm beer bottle. I had another shock when I saw red seeping on my palm, but it was just the red ink of the bottle's label, running with the sweat. Then I did get up, and went inside. I was hungry, I needed water. I suspect I was seriously dehydrated. Inside, the clanking fan flung heavy air around the room. A last shred of light picked out the empty bed. I'd left my towel there, and when I picked it up it was warm and wet, as if taken damp from a dryer.

There was a bottle of water in the fridge. I drank from it in gulps as I smeared peanut butter on stale bread. I was glad it was too dark for me to see if the bread was moldy. I found also some kind of healthy rice cakes that Sylvia's father had given her, extolling their excellence. They tasted like crunchy air. I covered two of those with peanut butter and ate them.

Of course I could have left Sylvia's room, could have gone home, but the thought never occurred to me. I was waiting here for Sylvia, that was what I had decided to do, that was what I was doing. Otherwise, I would have had nothing to do. Maybe there was another reason I chose to stay in that room of many discomforts. I felt like a prisoner. Or I felt like an ascetic monk, since I had chosen. I did not think much about where Sylvia was. I only thought of her arriving here. Of course I knew where she had gone. But I didn't. I was sure that if I had driven to Carla's I would have found her there. They were up on the roof of Carla's building, drinking gin & tonics. Carla had said once that if she knew the end of the world was coming, the missiles were on their way, she'd go up on her roof with a pitcher of gin & tonics and raise a glass to the ICBMs as they tumbled down from the sky. I had never seen a pitcher of gin & tonics; maybe she'd gotten the idea from a pitcher of martinis. Sylvia and Carla now were drinking gin & tonics, and talking and laughing, and not worrying at all about Sylvia's mother, or grandmother, or grandfather or aunts and uncles or cousins. Sylvia was just glad to be away from them all, and with her good friend, and had no qualms about forgetting all of them. Why should she?

Sylvia and Carla were up on Carla's roof, drinking gin & tonics, talking and laughing. This certainty, phantom but sound, impressed itself on my mind. It did not disturb me, because it was such a pleasant scene. Maybe the rooftop was empty, but Sylvia and Carla sat there, drinking gin & tonics, talking and laughing, and watching the fireworks. The distant popping sounds did not draw my attention at first. Only when I noticed a purple starburst above the trees did I remember that it was the Fourth of July.

I went out on the landing and leaned on the railing. I could see fireworks of varying intensity from all three directions visible to me. The brightest were to my left, the west, from a park on the river near downtown. Those would be the ones Sylvia and Carla were watching. They would have an excellent view. I looked for a while. I could see them quite clearly from here. The muffled explosions reached me much delayed behind the bursts of color. That was the only sound. Because the sound came so far behind, and so distantly, it felt like every burst was a dud. And then, all of a sudden, a wash of sadness swept me, it seemed so utterly pathetic to be watching fireworks alone, it was unnatural, unspeakable. What creature is so miserable, so despised, so unfit, that he must watch fireworks alone? The whole city was gathered to watch the fireworks, to ooh and aah and honk their horns, and everyone knew that I was watching fireworks alone, and I was the only one. Everyone knew that my girlfriend was watching the fireworks with her girlfriend. I remembered my mother telling me that the first time I was taken to see fireworks, when I was just two or three, I was so frightened by the noise that I started wailing with the first blast and wouldn't stop. I clung to my father's neck and screeched in his ear. People around us complained. There was nearly a fight, prevented only by the fact that with me clutching his neck, my father couldn't throw a punch. Finally we had to leave, and my brother was so angry that he didn't speak to me for days. Now I was highly susceptible to self-pity, I succumbed, I let the cheap sentiment of the memory reach me. I leaned on the railing, near tears. I sent out a silent plea to Sylvia to please, please, appear now in the streetlight down below, and climb these stairs to me. I wanted to cry, and yet I knew that if I started I wouldn't be able to stop, I was so exhausted, so wrung by the day's tensions, the sun, overexertion, beer, the heat. I forced down the tears. I thought also that I might be sick. I went inside.

I had never slept in Sylvia's bed alone. I couldn't do it now. I took the cotton blanket from the bed and spread it on the braided rug. I laid my head down and realized I'd forgotten a pillow, but before I could reach for it I was asleep.

. . .

I woke just before the rain. I stepped out into the still drench of darkness. Fireflies blinked blurry in the mist below, like ships in the distance. A firefly night, a firefly night—it seemed I had heard the phrase somewhere, I couldn't remember where—out into the firefly night. The whole world quivers with beauty and meaning. It's a pity we don't understand what it tries to tell us. I lit a cigarette and remembered where I was, and who wasn't here. I had thought I would go to sleep, and wake to Sylvia's touch; I had gone to sleep on this faith, like a child. And now that she wasn't here, I had to believe it. The sick despair swelled back. Again a phrase of Sylvia's: the unimaginable had collided with the undeniable.

The air was somewhat cooler, still uncomfortably warm. And then there was a change in the air, strange and frightening. It swelled somehow, thickened in my mouth and lungs. It moved toward some critical mass, the point of saturation—saturation of moisture, scent, omen. I could smell everything, all at once—earth, grass clippings, sweat, trash, hot tar, cats, moldy bread, rubber tires, stale beer, newsprint, cooking grease, car oil, piss, rust. It was hard to breathe—more, I didn't want to take a breath. The change in the air was like a rising scream,and it went on to the point where had it continued an instant longer, I couldn't have endured it, no one could have, the city would be snuffed.

Then rain. Like in the movies. From the enveloping stillness, a gash of lightning, snap of thunder, rain. A cloudburst. Straight down, and seeming solid. As the first of the rain hit the roofs, the ground, the streets, the intensity of smell peaked, then vanished. I felt as if I'd been lifted in the air by a crushing fist, then flung down. In seconds I was drenched. I dropped the sodden cigarette from my fingers. I thought I ought to enjoy this sudden relief, release, thought I should raise my face to the rain in praise. I stood another thirty seconds thinking this, and trying to enjoy it. But I couldn't. I was cold, I began to shiver. I hesitated another moment, then turned and ducked through the doorway, feeling defective and ungrateful.

I peeled off the wet shorts and rubbed my head with the still-damp towel. I lay down on the rug and wrapped myself in the cotton blanket. It was still hot in the room, there was no wind. I felt cold and hot and clammy. I put my head down as sleep swept in, then started up at the sound of stones thrown against the windows. Sylvia was throwing stones at the windows. No, that was stupid. But stupid things could be true. But mere hope wouldn't make them true. A pinging clamor on the landing. Hail. White pellets of ice, the size of mothballs, some big enough to lodge in the grating. Propped on my elbows, I watched as the hailstones melted and slipped through the grating. There and then gone, there and then gone. Sleepily I tried to watch the several stones and see them as they fell, but my eye was always absent when they slipped through, and then others took their places. I lay down again. Hail the size of peas, size of mothballs, tennis balls. Some places they had hail the size of softballs, size of grapefruit, size of bowling balls. Places like Kansas, Oklahoma, Nebraska. *Believe It or Not* books said frogs, seashells, toasters sometimes rained down on peasants who spoke inscrutable dialects. Out on the faceless plains, dire wonders fell from the sky. No one you knew ever saw it.

Icebergs

This was a room where you could write a book. The book would be speculative, resigned. The book would be reaching, and always thwarted. The book would speak of beyond but return always to here. The book would be a sad book. Would it have to be? Yes. You could write a happy book in this room, but it wouldn't have been written in this room.

Sylvia wouldn't write a book. Sylvia wrote in her journal, about such things as a bed of gray ice. She wouldn't write a book. She didn't write poems anymore.

Sylvia had tacked up pictures on the wall, pictures sent from an imaginary friend and she was the imaginary friend. Sylvia hated paradoxes. The pictures showed snowstorms, thunderstorms, tornadoes, the city locked in bitter cold, rimed in ice.

Then was the city for Sylvia defined in extremity and crisis? That was the weather that made the papers. But she might have put up postcards of a placid, clean, industrious place, sunsets and still lakes. She might have put up nothing. Now there was nothing. Only the Christ and his disciples, unworthy, every one. What were they eating? The food looked unappetizing. They were eating the Christ. Only a mirror. Only a window.

A molecule on the shelf. It was something or nothing. It was matter that was or that could be or couldn't. It was an exploding sun. It was the cause of the heat.

Did I really wake with the smell of wet scorched earth in my nostrils, its taste in my mouth? The rain would have snuffed the grassfires burning around the city. But maybe it was just having smoked too many cigarettes, or the smell of an overfull ashtray in the damp air. As soon as I woke I got up, put on my grimy damp clothes, and left. I may not be believed if I say that I again forgot about the car, as after my first night in the room with Sylvia, but it was so. And again I found it there on the street, but seeing it and remembering it elicited no particular reaction. I just got in and started it up and drove away. I drove along the river and through downtown, past the basilica and along the parkway to the lake, past my mother's house. Now I was consciously repeating my course from the morning after my first night at Sylvia's, to no particular purpose. A pattern had emerged and I followed it. The pattern was arbitrary, a chance sequence. If you say two, four, six, then you must say eight. I was just saying eight. I didn't imagine that it meant anything. It was just to give it reason. Just to say this is why.

I stopped by the small round lake in the center of the chain of lakes. It was the most open, the noisiest, the least interesting of the lakes. I got out of the car and walked across the bicycle and walking paths and down a short weedy bank to the shore. The hot dry summer had shrunk the lake, creating a narrow strip of beach here where usually there was none. Although the rain had freshened the air, it hadn't been enough. The heat was returning, and the smoky haze. It didn't feel like morning, but only the end of a long hot night. The narrow beach was rocky, and covered in a gray leach of scum. There were a lot of dead fish on the shore, little panfish and minnows, and one big dead carp. The lake was green from the early algae bloom, and it smelled of vegetable rot. The fish on the shore added their fleshy stink, and there was the tang of smoke in the moist air, too. I wondered why I was walking here. Because it's nice to walk by the lake in the early morning. I saw a man on the walking path ahead, coming my way. He had a dog with him, like a husky but I think it was a mutt. The dog was unleashed and it came down to the scummy shore and sniffed and then flopped down for a happy, vigorous roll. The man yelled:

—Oh for christ's—get the fuck out of—King, come here, *heel*—son of a—

The dog hopped to its feet, bright-eyed, and leapt up the bank and ran to its master, an old guy. The dog bounded happily because it's nice to take a walk by the lake early in the morning. The dog bounced to the man's side for affectionate reward. The man was repulsed. He was carrying a leash and now he bent over and tried to attach the leash without touching the dog. He recoiled several times from the dog which shivered with delight, and then he finally got it, and they continued along the path. I heard the man say:

—Okay, King, that's your fun for the day. Your mother's putting you in the bath when we get home.

The dog's mother was the man's wife. I liked the dog and I liked the man, too. You're supposed to keep your dog on a leash, and clean up after it if it shits. I made a U-turn up the bank and followed behind them on the walking path till I reached the car. I left the stinking lake and drove to Frank's building. I didn't think Frank would show up at my door today and take me swimming, so I went to him.

Frank's door was not locked and I went right in. He was sitting at a drafting table drawing. There were cartoons on TV. He said:

—Jesus christ, did you spend the night in the gutter?

—Almost, I said.

—Want some coffee?

—No, water, just water.

The heat was returning already. A feeling as of thick wet wool. I said to Frank:

—Did it make it to a hundred yesterday?

—I don't know. It was fucking hot. Did you watch fireworks? I tried to call you.

—Yes. But, no, not much. Did you, have you seen Carla?

—No.

It was the roadrunner on TV. Clouds of dust in the desert.

—I can't take this heat anymore, I said. I can't take another day.

—It's cool on the lake.

—What lake?

—What lake. The big lake. *Gitchee-gumee,* the big sky water.

—Let's go there then.

—You want to go there?

—I'm going. Want to come?

—Yeah.

I pulled myself up in the stuffed chair where I was slouched.

—We'll pitch a tent on the big lake. We'll go naked. We'll live like animals. We'll never come back.

—Sounds like a plan.

—What are we waiting for?

Frank grabbed a watercolor pad and some implements of art. He stood up and said:

—I'm ready.

He was wearing cutoff jeans and a tank top.

—It's cold on the lake, I said. Get some pants, a sweater, your sleeping bag. We're going camping.

—Are we really?

—Yes.

I helped him gather his things. We drove to my building and I sent Frank into the basement to find my box of camping gear while I went up to shower and change and pack some clothes. We stopped for groceries, sunscreen, beer and whiskey. Then we set off north through the steaming yellow heat.

We passed close to Gran's house on the way out of town. Then through the willowed sand flats, and then rolling country of hardwoods, into which birch and pine gradually impinged. We drove through granite roadcuts and the farms dwindled. We stopped for breakfast at a truckstop and discussed where we should go. We settled on stopping as soon as it got cool. The point wasn't to drive, or to see the most beautiful places. The point was to be by the big lake where it was cool.

It was just before noon when we crested the height of land and the lake and the north shore and the port city spread out below us. The islanded delta of a beautiful river fanned out into the harbor. Topping the height of land here and seeing this great spread of water had always brought me a swell of exhilaration, a sense of joy and boundless possibility. Even today I did smile. A small edge of hope touched my emptiness. I drove too fast down the long steep hill into the city, hurtling past down-shifting trucks, their air brakes screaming. Our ears popped on the descent and we felt the change in the air already. We were entering another atmosphere; it was as if we had crossed into another country. The route to the north shore traversed the hill on which the city was built. The harbor lay on our right, and to our left the streets ascended sharply. The way out of town took us down an elm-draped street past the lawns of great mansions that fronted the lake. Money from lumber and iron and ships. We crossed steep streams that jostled over rocks. Then the road emerged from the civil foreign elms, and gave us the big lake, rock shore, aspen and pine. I pulled in to the big gravel lot of a tourist information site to use the bathroom and to put on more clothes—we were already uncomfortably cold in our shorts and T-shirts.

I parked next to a big Lincoln. A gray-haired couple stood in front of it, and as I climbed out of the car I heard the woman say:

—It looks just like *Maine,* doesn't it?

She was snapping pictures. The man said:

—Where's the salt spray? Where's the seaweed?

—I didn't say it *smelled* like Maine, I said it *looked* like Maine.

—Where's the lobsters, clams?

—It *looks* like Maine. There's *fish.*

—Fish. It isn't Maine.

—I didn't say it *was* Maine, I said it *looked.* . . .

When I came back from the bathroom the Lincoln was gone and a van had pulled into its spot. The doors rumbled open and a half-dozen women wearing black rushed out. I thought of a circus car. But they were nuns, in full black habits and cowls over their heads. They were middle-aged and they almost sprinted from the parking lot toward the shore, nimbly surmounting an iron guardrail at the lot's edge. A strong wind off the lake spread and flapped the loose cloth of their habits like poorly trimmed sails. They were speaking a foreign language that I thought might be French, but then I heard an older woman in a gray suit, the last to exit the van, address the driver in German. I had lost track of Frank, and I walked toward the shore to look for him. The shore was a solid sheet of rock, dark gray basalt encrusted with quartz. The surf came in heavily on a northeast wind. I saw Frank crouched down near the water. A little distance from him the nuns were hopping and laughing and speaking fast high-pitched German that swirled in the wind. They clutched each other's clothes as they danced at the edge of the waves. When a wave crashed toward their feet they ran back, and their cries of delight swelled. If one went in here she would be sucked out into deep water in an instant, and wrapped in icy voluminous black, and drowned. I wondered if they were crazy, if they were on a field trip from a home for mentally defective German nuns. But it was fun watching them. I called to Frank, and had to shout several times before he heard me over the wind and waves.

Traffic was heavy on the north shore road, cars full of kids, sightseers, cars pulling boats and campers, RVs with wooden plaques on the rear informing us that this was Betty and Irv from Dubuque who were making us crazy, taking the curves at 20 mph. It was a hilly, winding road, two lanes, dangerous passing, and anyway there were slow-moving vehicles as far as we could see. We ate chips and drank pop and smoked cigarettes. Frank tried to find a station on the radio, but the cliffs and hills made reception iffy. In about an hour we reached the entrance to a state park on the lake, but what should have signaled a joyous journey's end was not a happy sight at all.

Cars and campers were backed up into the highway's turn lane and the line of vehicles led two hundred yards up the park entrance road to the little *faux*-rustic reception house where the rangers registered campers. And the rustic sign said: CAMPGROUND FULL. Betty and Irv had planned on camping here too. I slowed and stopped behind their RV. We both stared silently at the line of idling vehicles.

—Why are they all waiting here if the campground is full? I said.

—Why are we waiting here if the campground is full? Frank said.

—Oh god. I don't know. Let's just go home.

—No no. Just wait here. I'll go see. I'll go talk to the Mr Ranger Man.

Frank left and I sat feeling sick. Cars whizzed by, rocking the Bel Air with their wind. Kids dashed among the cars in this little wilderness traffic jam. All these idiots had their cars idling, though no one was moving, and the air reeked of exhaust fumes. And it was hot here. The lake was maybe half a mile away over a hill, and the difference was twenty degrees. I pulled off my sweatshirt and waited. Frank returned in just a few minutes. He came loping down the hill and got into the car.

—They're all waiting here in case some campers leave, he said. Checkout time is two.

—Checkout time. Great.

—But there are some backpack sites down the shore. Rather primitive.

—Primitive is what we want.

—They're about two miles from here, the ranger says.

—Two miles. We don't have backpacks.

—You're not really supposed to, the ranger says, but we can park on the frontage road that used to be the old highway, and from there it's just a quarter-mile or so.

—The ranger says.

—It's the road the fishermen use to get down to the bay.

—The ranger says.

—The ranger says that in order to fully enjoy these back-pack sites you ought to have the arduous experience of walking two miles to get there, but he really doesn't give a shit.

—God bless the ranger.

—We can take the site and come back tomorrow and pay.

—Quoth the ranger.

I spun a U, feeling like a yahoo, and drove at high speed back down the road, missed the turn the first time, made another reckless U-turn, jounced off the highway onto a stretch of cracked weedy blacktop, and parked in the brush that was encroaching on the road.

—Here we are, said Frank.

—Where do we go from here? I said.

—I have a map.

We carried the cooler between us, with the box of camping gear on top, and loose items tucked under our free arms. The box fell off a number of times, and we kept dropping things. But it was an easy walk to the campsite, following an unmaintained section of the old highway most of the way. It felt as if we were on the path of a dead civilization: the dashed yellow line had gone ghostly faint under weather, the old blacktop had erupted from frost-heave, and trees and weeds pushed up through the cracks and pressed in at the road's crumbling edges. But the sound of a steady stream of traffic on the main highway still came clearly through the trees. A hiking trail crossed the old highway. Frank said turn here, and we did, and within five minutes we had reached our campsite and the lake. Tall aspen ringed the campsite on three sides. On the lake side it was open, giving us a view of a small bay and a curving stone beach. To our left the shoreline rose to high sheer cliffs, pine-topped. At the other horseshoe end of the beach there were also cliffs, more modest. And the lake went out blue, blue, and blended with sky. It looked like Maine. The campsite consisted of an iron fire ring, a cleared dirt tent pad, and a charming little bench / table arrangement of two leveled stumps with a plank nailed across for seating and a rough countertop of a wide split log fastened between two aspens. You sat at it just as at a lunch counter, and we were soon calling it the Little Espresso Bar in the Woods. The toilet was like an outhouse without the house, a stool over a hole a little distance off and secluded by the brush. It took only a few minutes to set up camp. My tent was a slick little dome thing that went up in a snap. We threw our sleeping bags and pads inside. Frank had brought his pillow, and I gave him shit for that, but I wished I'd brought mine too.

With our camp set up we had nothing to do but amuse ourselves. We took beers from the cooler and went down to the beach, Frank with his sketchpad and I with a book. The beach here in the bay was formed entirely of stones, ranging in size from pebbles to boulders, the various sizes residing with their like in distinct sections, or strips, from small to large moving away from the water, a natural supermarket of stones. We found a patch of small smooth stones, half-dollar size, below a gathering of boulders. The stones were warm and made a lovely chinking sound as we settled into them. The high cliffs to the northeast broke the wind, and the bay was calm, and it was warm in our little stony hollow in the sun. I smeared high-proof sunscreen on every exposed inch of skin—my shoulders still pulsed with heat, and I thought I could feel the skin rising to blister. I started reading, but soon lay back with the book across my chest and fell asleep.

When I woke up Frank was not beside me. I looked around and saw him floating in the lake a short distance off shore. I yelled:

—Isn't it cold?

—Yes, he said.

I went down to the water's edge and tested with my foot. It was cold, but not so cold as I expected. It had been an unusually warm summer, and the bay, I guessed, was relatively shallow.

—Okay, I said. I'm coming in.

I pulled off my shirt. The first touch of sun on my burned shoulders felt caustic.

—Did you bring a suit? I yelled to Frank.

—You said we were going to go naked, live like animals, Frank said.

—Right.

I stepped out of my boxers and tossed them on the stones with the rest of my clothes. I inched with creeping steps into the water. Ankle deep, mid-calf.

—Careful, Frank said. It's slippery.

The bottom was stones the size of heads, all covered with a thin layer of dark algae. The water was almost numbing cold. By the time I had waded to mid-thigh my scrotum had tightened in dread anticipation—all mechanisms, save my brain, rallied to ensure posterity. My foot found a smooth slick stone and anticipation ended. I had wanted ice in my organs; now I had it.

—Oh! I said. Oh! God!

—Is it good? Frank said.

I shook water from my hair.

—Yeah, I said. Yeah, it's okay. I don't think I can stay in long though. I don't think you can really get used to it.

—I've been in for twenty minutes or so.

—Yeah but you're cold-blooded. Fish blood.

I swam out a ways. I did get used to it a little, though I could feel the chill continuing to penetrate. The water was so clear, when I opened my eyes underwater I could see the roundstone bottom sloping down a long way into darkness, and looking up, the water was a wavy window on the blue sky and the yellow shape of the sun.

I returned to the beach, groping on all fours up the slippery rocks at the shore. Frank sat naked in our hollow. I pulled on my T-shirt and joined him there, replenished my sunscreen, and we warmed ourselves in the stones and the sun. I rolled over and lay bare-assed to the sun while I read. It felt ridiculous, and it felt good. We spent the afternoon going in and out of the water, cooling and warming. We saw no one but one young couple who came hiking around the northeastern point, and we saw them in time to make ourselves decent. When they passed us we smiled and waved, and they said, nice spot you've got, and we said we liked it fine. Maybe others approached when we weren't looking, and fled at the sight of us. But we were undisturbed the whole afternoon. Small waves chimed the wafered stones at the water's edge. We slept and woke, swam and sunned. Frank went gathering agates from pebble beaches, and returned with a handful glowing rich brownish red.

—They're easy to find, if you just free your eye and look for light instead of rocks, he said.

I tried it for a while, and found a few, but I was not a virtuoso like Frank of the liberated eyeball.

—Maybe this is what it would be like if you were a little tiny creature on a regular sand beach, Frank said another time. It looks smooth and flat to us, but there's really an extremely various and intriguing landscape there. The sand looks packed but it's full of air, and there are rises and dips and hollows and crags. There are worlds within worlds, my boy. Nothing is quite solid or quite as it seems.

—Bodies never come in contact, and the soul never touches its object. An innavigable sea washes with silent waves between us and the things we aim at and converse with.

—Quoth the ranger?

—Quoth Boscovitch, via Emerson.

Neither of us knew anything about this venerable Boscovitch whom Emerson had cited offhandedly. He was some faddish forgotten physicist, we guessed. We began referring to the ranger as Boscovitch, a man of mystery and wisdom. He puzzled out all of life's conundrums for us as he registered campers and sold firewood by the bundle.

We lost the sun earlier than we'd expected behind the height of land. The air cooled quickly and we were still exposed on the beach as the sun's warm burnish turned to gooseflesh. We climbed through the stones back to camp, put on jeans and sweatshirts, and scavenged for wood to build a fire. The site's previous tenants had left some seasoned wood, and we used some of that to start a fire and dry the half-damp mossy branches we found in the woods around. Supper was canned beans and bratwurst grilled on sticks. A light breeze off the lake was enough to keep the mosquitoes away.

When the last light faded there appeared overhead such a sprawl of stars it was both breathtaking and garish to us who saw only the dimly spangled city sky.

—It's not the same world, Frank said. It's just not the same at all. The sky in the city presses down, it hems you in, and here, it draws you out and out. How do we allow these things to happen? How have we let the sky be taken from us?

—You want to move into the country? I said.

—No. It scares the shit out of me.

—What scares you?

—Bears, for one thing. Boscovitch said there were bears.

—Of course there are bears. This is the woods. This is where they shit.

We laughed. We were drinking whiskey. Frank said:

—No, I mean there are bears in the campground. It's been so dry, the woodland creatures are suffering. The bears are ransacking the campgrounds. Boscovitch said don't leave any food outside. Don't put anything tasty in your tent, even toothpaste.

—Now you tell me. We'll have to hang the cooler in a tree.

—Hang it in a tree.

—Yes.

Intoxicated as we were by sun, water, wind, and whiskey, the complexities of hanging a cooler in a tree were long and engrossing. The box of camping gear yielded a bungee cord with which we strapped the cooler shut, and line enough to reach a stout enough limb in one of the aspens. We were inordinately pleased with ourselves when, after many failed attempts, we managed to toss the line, weighted with a stone, over the limb, and hoist the cooler aloft. There it tilted and yawed in the breeze, and it made me think of flotsam in the trees in the aftermath of a great flood. We sat back down and clinked our tin cups in a toast to our ingenuity.

—Won't Boscovitch be proud of us, I said.

—Boscovitch is beyond pride, Frank said. We have done right and well, and that knowledge must suffice.

—Well then fuck Boscovitch, I said.

We sipped the whiskey mixed with lake water. In turn we wandered away from the fire, burning low, to look at the sky. We looked out onto the lake. The lights of a freighter appeared far out on the water. Satellites passed with remarkable frequency overhead, and shooting stars streaked and vanished. If the sky was more brilliantly starred here, so was the background darker, a pitch volume of awesome depth. The lake too was a giant bowl of icy black, scooped then filled by glacial scour and melt. Watching the freighter, I shivered. I imagined standing its steel deck and looking out in sight of nothing but dark cold water, and looking down at the ship's swirling crease on the face of bottomless black. I imagined icebergs; there were no icebergs. What could take a fit ship down on a calm night? The simple draw of darkness. It made no sense, and that was just it. The dark of the lake, and the sky, and of the woods that from here ran north to tundra, to the even colder waters of the Hudson Bay, millions of acres so dark, so sparsely starred. It was a wilderness. What wasn't? The dark here all around was the dark of unreason, and of separation, and of the soul's inane, faulty, aimless conversation. The dark of blood, dark of the tribe, of the cave. We had left the cave and had no home. We knew too much and understood too little. The forest animals were at home in the wooded dark, and the stars in cold empty space, and the fish swam unblinking in icy blackness. And we cowered in a wilderness of unreason and caprice and random cruelty. Idiots everywhere we were. So then it seemed. Space was the dark clot of a brain massed with ganglia and neurons that fired without cause. A splendid and repulsive spectacle. Thus again the tribe, the community of blood, defended to the death against all others. All the tribes called themselves the people, thus the other was not. An innavigable sea. Its silent waves. *Innavigable.* It meant you couldn't get across. But innavigable waters had been navigated before. By fools who if they lived were heroes. Then innavigable meant that you shouldn't try.

Frank came back from a look at the stars and sat on the log beside me.

—What's the matter? he said.

I leaned forward clutching my elbows and shuddering slightly.

—I'm cold, I said.

—Before you were too hot, now you're cold. Aren't you ever happy?

—Innavigable, I said, slurring, as one would tend to do even under ideal circumstances.

—Innavigable, said Frank. Do you think we need to worry about those bears?

—It's usually the small ones that come into camp.

—I reckon any bear is big enough.

—What good is worrying?

—You eliminate possibilities by considering them. If you were able to consider every possibility, nothing would ever happen again.

—Wouldn't that be fun.

—You'd bring time itself grinding to a halt.

—Let's give it a try.

—Yet God must have considered all possibilities, and yet time continues, ergo there is no God.

—You're being logical again.

—It seemed appropriate. You're acting spooky.

—I'm scared.

—Of the bears?

—No.

—Of the dark?

—Not exactly.

I lit a cigarette with an ember from the fire.

—God has considered everything, I said. Time does not exist, it's our misapprehension. Possibilities appear as God considers them, then vanish. Everything's already over, we just don't know it. Time doesn't go forward; there's a great mass of chronicity in the mind of God, and it's done. If you were the tiny creature looking up from the sandy beach, what do you think you'd see? This cauldron of time and the mind of God are too vast for us to comprehend.

—We can't even see to Wisconsin, Frank said, pointing out across the lake.

—Quite so.

—Then you're saying you believe God does exist.

—Probably not.

We saved out some firewood for the morning, and burned what remained of the rotty sticks and branches we'd gathered, and went into the tent. I was drunk enough that I went right to sleep, bears or no. My sleep was troubled with dreams I didn't remember. I only guessed they were about deadly icebergs on the dark sea of unreason where I sailed lashed to the mast of a doomed ship. I only surmised a howling in the cave, faintly echoing as I woke.

I woke in grayness beside Frank's empty sleeping bag. I ducked out of the tent into fog and chill. I could not even see the lake from the campsite. When I reached the shore I heard a rhythmic liquid shooshing, and saw a curling wake that marbled the still water, and followed it to find Frank at its head. I watched him for a minute without speaking, then returned to the campsite and built a small fire.

Frank was glad for the fire when he returned from the beach. I had cooked up some boiled coffee, exceedingly strong, and we drank it heavily sugared. We cooked bacon draped over sticks, the way we'd seen it done on TV or in a movie, and didn't bother with the eggs. You could get pretty uncivilized pretty fast. As we finished eating the air began to move with a wind off the lake, and it steadily gathered momentum, and we went to the edge of the woods and looked toward the lake and stood in the rushing fog feeling as if we were moving at great speed ourselves. We both sat down on the rocks at once. The fog thinned, filtered lights glinted silver on the water, the cliffs appeared in dark outline, the sun a hazed glare over the lake, and the focus of everything sharpened, sharpened, sharp, and we were sitting in bright sun under blue sky. We turned and saw the mass of fog shredding through the lakeside trees, fleeing and dispersing like a routed army up the height of land. All of a sudden it was warm. Our faces and hair were wet. Our damp clothing steamed, as if we were composed partly of fog, a component of darkness which the sun drove off. I went toward the water shedding clothes as I walked, and waded in for my first swim of the day.

We settled in again to the routine we'd established the first day. Frank went about mixing pigments from red and blue berries, charcoal, green berries and leaf, rusty orange lichen. He had every available cup, bowl, and can arrayed on the Little Espresso Bar in the Woods, and he mashed and mixed like a primitive alchemist. Even the coffee grounds were put to use. I just lay in the sun and dipped in the lake. Once I walked down the shore a little ways toward the lower cliffs. I sat a while on a basalt outcrop that dropped steeply into the water, and looked down into the rolling, swelling mass of liquid glass. The sun reached deep, lit the water as if from within. The night, the images of pitiless dark water, they were gone, like a dream. I knew the deep waters were a danger even by day; but here on the shore the lake was soothing, hypnotic. Rock and water, sun and sky, all elements met here in ordained intercourse, and it was easy to believe, with warm stone on my back, my feet cool in the water, light warm wind on my face, that everything belonged to one splendidly textured whole. That everything truly was as it was meant to be, and our understanding was joyously irrelevant.

. . .

Frank and I were in the water, complacent in our aloneness, when a shout reached us from the shore.

—Hey, you guys!

My heart jumped and I remembered I was naked. My clothes were on the beach and so was whoever had yelled. We blinked away water and fought the glare to see who had called. I made them out as they came out of the shade onto the sunny beach. Two young women with backpacks.

—Hey, are you guys keeping this site tonight?

The one who spoke had a bob of wavy red hair, a T-shirt, baggy khaki shorts, heavy hiking boots. She leaned toward us with one foot raised and propped on a rock. She pushed hair back from her eyes. The other, taller, dark-haired, stood behind her.

—Are you staying here tonight?

Frank and I looked at each other as if to decide who should answer. Finally I said:

—Yeah, we're staying. Probably just tonight.

—The site marker is empty, did you register? did you pay? All the other sites are full, the redhead said.

—We're going to register today. It was too crazy yesterday.

—Boscovitch said it was okay, Frank said.

—What? Who?

—The ranger, I said. The ranger said we could.

The redhead turned back to the dark-haired one, and whatever they said we couldn't hear. They looked around the beach and maybe they saw my book and Frank's sketchpad. The redhead made a gesture and the dark-haired one gave a sort of nodding shrug.

—Do you mind if we share the site? Red said. We'll pay for it. We'll go back and register.

Now Frank and I conferred and agreed. Why not.

—Sure, I said. If you can find a place to pitch your tent.

—We'll just throw our bags down, Red said. The bugs shouldn't be too bad. Thanks.

They went and dropped their packs at the campsite, and came back to the beach. There was no time to make a dash for clothes. Frank was floating happily on his back, as if our predicament hadn't occurred to him, or maybe he didn't see it as a predicament. The two women sat in the rocks and took off their boots. They talked together and laughed. Now they were close enough that we could converse more or less normally. The dark-haired one said:

—Isn't it cold?

—Pretty cold, I said. It's good though.

Red came to the edge and tested with her toe.

—Yow, she said.

—Yeah, I said. I was thinking about getting out, but, I don't have anything on.

They laughed.

—I can see that, Red said.

I didn't think she could, because though the water was clear, the glare and reflection must have covered me.

—It's okay, you can come out. We'll avert our eyes.

They moved down the beach a little, and did look the other way, and I came out of the water, not hurrying, because you couldn't hurry on the slick stones, and once I was standing on the beach I felt a bit emboldened, and took my time drying off and dressing. The women came back down the beach, and we talked. They went to college at the state school in the port city. I don't remember what they were studying. Red had perfectly Irish features, small upturned nose, copious freckles on her fair face and arms. She was smaller than she had first appeared, in her big boots with her foot on the rock and the big pack on her back. She was outgoing and funny. The dark one was her opposite in every way. She seemed shy, didn't say much except when Red spoke to her.

Frank paddled and floated, oblivious to all of us. We watched him. When he began swimming outward in a breast stroke his white ass breached the surface, and we all laughed.

—Your friend is a good swimmer, the dark one said.

—Yeah, he's a water baby.

Red turned to her friend and whispered something, encountered opposition, persisted, and received agreement, and turned back to me.

—We're going to go in. We don't have suits either. Be a gentleman, will you?

—Always, I said.

I took my book into our hollow, and propped myself up with my back to them. I heard them scuffling in the stones and giggling, then the march of their bare feet down to the water, and whoops and cries as they edged in. I found my heart beating inexplicably fast.

—Hey, you, swimmer, cut it out, turn around! Red yelled.

I looked out in time to see Frank disappear beneath the surface like a loon. Like a loon he was down a long time and eventually surfaced someplace unexpected. I saw also the bare backs of the women as they stood in water just above their waists. The dark one stood with her arms crossed on her chest. Red threw back her arms and pushed out into deep water with a yell. When she surfaced and caught her breath she began to tease and cajole the dark one till she too floundered out and ducked under with a moan.

The two bobbed about for a while, remarking mostly in gasps on how cold the water was. Frank kept his distance from them, darting and diving, approaching and withdrawing; his actions were remarkably like a loon's, perhaps he was thinking like a loon. He was teasing the women, there was some kind of tension between them. Frank's feigned obliviousness, when in fact he was acutely aware, was part of his intuitive method, I guessed. Shortly the women came toward the shore, Red calling:

—We're coming out.

I turned discreetly away. The women said *brrrr* as they put on their clothes, and then they came and sat down with satisfied groans in the warm stones near me. In a few minutes Frank appeared in his stealthy way at the shore, and strode dripping up the beach toward us. The women turned away, hid their eyes, clutched each other and laughed. I was embarrassed, I threw Frank his clothes, encouraging him to dress if he hadn't planned on it. Frank took things on wholeheartedly. I had said we'd go naked and live like animals, and he had taken me at my word. When Frank was dressed then the introductions were made. The dark one was Maria and Red was Katherine; Maria addressed her with a bewildering barrage of nicknames—Kathy, Kate, Kat, Katie, Kath, Katerina, Kiki, KayKay, and just plain Kay or K. They said they had planned on doing some backpacking, but now that they were here, they thought they would stay. They liked our setup. Frank asked if they had hiked in from the main campground.

—No, we parked on the service road, K said. Is that your car up there, the old blue thing?

I hesitated.

—Yeah, sort of.

—What did you, steal it?

—Not exactly.

—My grandpa had a car like that, K said.

—We thought it was a junker just left there, Maria said, laughing.

—That car is someone's grandma's car, Frank said.

—Whose grandma?

—A friend's, I said. But it really is my car, I think.

—You think, K said. I don't know about you guys.

We said we didn't know either. How could you tell, really? We didn't talk a lot, but it was comfortable, the four of us lazing in the sun. When we became too warm, or just when we felt like it, we went into the water. At first there was much preparation, announcement, averting of eyes, and nervousness. Then they moved themselves somewhat away, so they were not undressing right in front of us or we in front of them, and whoever wanted to swim just undressed and went in, and came out without ado. No one paid overt attention to this process. Which is not to say we were unaware. We all were aware. The tension gave a tingly edge to the sensual surround of sun, water, forest, sky, and stone. I wasn't entirely sure that Boscovitch would approve.

We all went up to the campsite for lunch, and shared out what we had—bread, cheese, chips, sliced meat, a smoked trout the women had bought at a fishhouse along the shore, oily, rich, and sweet. K brought out a bottle of rosé wine from her pack; I saw at least one more bottle, and I wondered how far they'd planned to hike with all that. I noticed also what looked like a corner of a brightly colored swimsuit. Since our every container was filled with Frank's pigment concoctions, he begged permission to drink from the bottle, and we all drank from the bottle, passing it around, and it was soon slick from our oily-fish-fingers. K took the bottle for the last swig, gripped the neck and tipped it to her lips as she threw her head back, then tossed the bottle over her shoulder with a laugh. The bottle was headed for soft grass, but it found a rock in there and shattered.

—Oops, said K.

We all helped gather the bits of broken glass. The bottom

was preserved in a jagged-rimmed cup, which Frank said he could use.

After lunch I headed back to the beach. Frank stayed at the campsite to play with his colors, and Maria, who said she did some painting, asked if she could join him. K came down to the beach.

Since we were two we sat closer together on the beach. Since we were two it was more uncomfortable undressing to go in the water. But not exactly uncomfortable. A couple of times we were in the lake together, and we bobbed nearby with only our heads exposed. Her eyes were pale blue and her fair skin now was pink. Her faint red lashes, when she blinked away the water, made her look young and vulnerable. As we treaded water together she said,

—So, Mr Big City Reporter, do you have a big city girlfriend?

I hesitated not only because of what had happened on the Fourth, but because it was hard to think of Sylvia as anyone's girlfriend.

—Yeah, sort of. I think.

—You sort of have a girlfriend you think. Like you sort of think that's your car up there?

—Exactly.

I left the water while K was still swimming, dressed and went up to the campsite, fighting the urge to look back. Frank and Maria sat on a log applying pigment to paper with their fingers. As I came closer I saw that they were actually using thin sheets of birchbark. They both looked exceedingly pleased. Each commented glowingly on the effects the other was achieving. I took a beer from our cooler and leaned on the Little Espresso Bar in the Woods.

K came up from the beach and stopped at the edge of the campsite.

—Have you been up around that point? she said, pointing at the tall cliffs at the mouth of the bay.

—We haven't been anywhere, I said.

—Well who wants to go?

—What for? said Frank.

—To see.

—See what? Frank said.

—See what's *there.*

—Much the same as here, I expect.

—Maybe you can see the lighthouse around the point.

—I'll pass, Frank said.

Maria said she would stay with Frank and they went back to pounding their berries in the tin cups and smearing it on the birchbark.

—What about you? K said to me.

—Sure, I said. Why not.

She put her socks and boots on and we set out along the shore. Just a few yards down the beach K stumbled in the stones and grabbed for me to stop her fall. Her hands landed hard on my shoulders. Reflexively I reached for her; I also cringed and shut my eyes against the pain and shuddered. She found her balance, removed her hands, then gently lifted the neck of my shirt to see my shoulders.

—God, you're sunburned. Sorry.

I sort of waved it off. I looked at her.

—You're looking pretty pink yourself, I said, turning away.

—Did you say I'd look pretty in pink?

I turned back to her.

—I said . . . Yes, you would look pretty in pink.

—I only go pink in the sun. Freckles and pink. Freckles everywhere.

The beach gave way to great tumbled-down chunks of rock as the shoreline steepened toward the cliffs. We found the trail that led that way, and walked along in the shade of the aspens. We talked little, and what we said was awkward and forced. Because we were thinking the same thing, or not thinking but feeling and anticipating. Something had been started. But we were strangers to each other. Probably nothing would happen at all. How could it?

The trail led upward, and when it neared the sheer rock palisade it turned inland. If we wanted to go up we would have to freelance it. There were small trails that headed up that way, and we started on one. It ended at a vertical rock face, and we went on by other ways that appeared as we came upon them, and they all ended in thickets of brush or cliffs. They were narrow and steep and dark in the shade of the trees' thick canopy. Sometimes we went down on all fours climbing slippery mossed stone, or went up hand over hand by vines or small trees that grew in sharp angles to the tilted earth. The last stretch forced us along a rock ledge only inches wide, moss-covered, with a scary drop below and dripping stone cold at our backs. We inched along, our backs flattened to the rocks. About halfway along I stopped to calm my nerves. K was behind me, and she reached for my hand, and held it, and we both gave a nervous laugh, and stood like paper dolls plastered to the rock. We could easily have fallen, and if we had fallen we would have been badly hurt. It was a stupid thing to do. When we reached the end and came out into the sun on the broad stone dome I was exhilarated. And my knees were shaking. We whooped and laughed and hugged, and said *God that was scary* and *Holy shit!* We were high, high above the water. The top of the palisade where we stood was like a great stone forehead, curving smoothly down and then plunging to the lake. A raft of mergansers down below appeared as specks on the gently rippled surface. Looking to the southeast I do believe we could see Wisconsin. It was thrilling to be up so high, and to have gotten there by such a dangerous route. We stood and looked and K put her hand lightly between my shoulder blades, and I rested my hand on her shoulder.

—We should plant a flag, she said. We should commemorate this mighty feat.

She turned and walked away from the edge. Just at its crest the dome broke up into a giant jumbled stairway, great squares of stone with trees and scrub and blueberry bushes growing in the cracks. She climbed up on one of these steps and sat down. I followed, and we sat on a cool bed of moss, in the shade of wind-battered scraggly pines, looking out on the lake. K began to unlace her boots, pulled them off, then pulled off her heavy wool socks. She chafed her ankles.

—God those things itch, she said. And those boots are heavy. Would you mind rubbing my feet for a minute?

I lifted her foot into my lap and rubbed it, and stroked her calf, the soft curving flesh. She lay back with her eyes closed and gave sounds of pleasure. I brought her other foot into my lap and as I began to rub it she sat up and put her arm around my neck and then we were kissing. And then we were kissing deeply and we were stretched on the bed of spongy damp moss. I reached under her T-shirt and touched her breast, cool and damp from sweat. She slid her hand down my belly and into my shorts, and her warm hand where I was still cold from the lake gave unspeakable delight. Her soft thigh too was cool and warmed under my hand. We had shed our T-shirts and then we were pushing off our shorts, and I said:

—Do you have anything? Are you on the . . .

—No.

—Then we shouldn't . . .

—We'll be careful. Stop before . . . I just finished my period, it'll be okay.

There was no question of shouldn't. I was on top of her and

I was in her. I was in her and I thought: *I am fucking her.* It wasn't making love or anything else. The language provides this specificity: *I am fucking her, she is fucking me, we are fucking.* Fucking on moss, on dirt, on pine needles, on rock. It could have been a romantic scene, lovers in the sun, in the wind, on the palisade head, but it wasn't because we were fucking. I knew I was doing a wrong thing. I was doing a wrong thing for no reason. I was doing a wrong thing for unreason. To immerse myself in it, to do hurt. And more, it felt right, insipid to say it, it felt *natural.* We had seen each other's bodies; we had been stroked and gentled by the textures of lake and earth and weather; to press warm flesh to flesh was to complete something. Even then I knew it was *wrong,* natural and wrong. But it required no effort or thought to be naked and entwined with a stranger in the moss on a rock. The press of flesh annihilates distance. Distance dwindling means distances exploding. Bodies never come in contact. The soul finds no end of desire. Good Boscovitch, we know.

She came, clutching me tight. She pushed me down and out, then pulled me back to her, and I came on her belly. I fell on her and the semen spread hot between us. I rolled away and immediately felt exposed and panicked. She lay back with her eyes closed, and then touched her fingertips to her sticky belly, and laughed. There were pine needles, and shreds of green moss, and dirt stuck to both our bellies. She sat up and looked at herself and said:

—How natural.

We wiped ourselves off as best we could with our hands, and painted the rock with it.

—That will have to do in place of a flag, she said.

She touched her chest that was red from the chafing of my stubble beard.

—Ouch, she said. You hurt me.

I was touching my raw shoulders. She leaned to look.

—But I hurt you too. Was it worth it?

The pain was searing and I felt sick.

—Yeah, I said. Of course.

Help me, Boscovitch: since bodies never come in contact, I never touched her, nothing happened, everything will be as it was.

I stood and brushed myself off. I was putting on my shorts when something rang clattering down the rock in front of us. Then again, and again. Someone was throwing stones. Stones on the window, hailstones. We heard voices, kids shouting. K laughed as she scrambled to dress. Three kids, two boys and a girl, clambered down the stairstep stones to our right and came out onto the bare dome. We heard their parents behind, yelling at the kids to stay away from the edge. The parents appeared shortly, a young, blond, athletic-looking couple. They saw us and waved hello, and we waved back. We were sitting nonchalantly. They didn't stay long. The parents were nervous about losing their kids over the cliff. When they left I stood up.

—Maybe we should go back, I said.

—I guess so, K said.

She reached up her hand and I took it and pulled her up. She kissed my cheek and I kissed her forehead. We went up through the rocks the way the family had come, to look for an easier way back.

We walked and talked and pretended that nothing had happened. We noticed things in the woods, wildflowers and odd fungi. As we neared the campsite I stopped and said:

—Maybe we should clean ourselves up a little.

We picked our way down the bank to the lake. The shore here was a shelf of dark basalt. I could see the beach in front of the campsite, and it was empty. I was hot from the hike and I stripped and dove into the cold water. As I went down toward the rock bottom it felt for an instant as if I were cleansed. K dove in, and when she came up she swam to me and held my arm and rubbed her hand on my belly, smiling. I did the same for her. We kissed, and then we climbed out of the water, dressed, and returned to the camp.

Frank and Maria were sitting at the Little Espresso Bar in the Woods, each drinking a beer. They had their paintings on the rough tabletop. When we came into the campsite Maria said:

—Hi. Did you see the lighthouse?

K looked at me.

—Well, no, I don't think so, she said. Did we see the lighthouse?

—No, I said. I don't think we did. We went up on the big cliff. I don't remember the lighthouse. We saw Wisconsin, I think.

—Yeah, we did, we saw Wisconsin, K said. It looked great.

It was late afternoon now. I didn't see how things could just continue. I didn't know what to do with myself, what to say. I remembered that we still hadn't registered with the ranger, and I seized on this small errand with enormous relief. When I said I was going to register the women said they would do it, they would pay. I said no, you're our guests. I insisted. I found my wallet and hiked toward the car.

As soon as I reached the old cracked blacktop I felt a thousand miles and a lifetime distant from everything. Because nothing that had happened in the last three days seemed to have any connection to my life. Saying eight. You can't just say eight. There might have been another sequence, not just two plus two plus two. Two four six. What next? Eight was the easy answer and not wrong but not the only one. I knew nothing about mathematics. It had nothing to do with mathematics.

Once in the car the temptation was great to drive and never stop. Canada was only a few hours away. I had a credit card. And I wished I were able to do that, just drive, and charge things. I was not. I turned into the park entrance, empty now and quiet, and pulled up by the reception house. As I turned off the car I realized I was now about to confront Boscovitch. I would have to confess everything, because he would know. He would stand broad-shouldered with his strong hands on his trim waist, and look straight into my soul with his deep all-knowing eyes beneath the brim of his wide crisp ranger's hat. But the question was, could he absolve me, or just throw me out of the park?

I opened the screen door and went in. Behind the counter with a map of the park under glass sat a woman with permed blond hair, like one of the cousins. I was confused, and when she asked if she could help me I almost asked for Boscovitch. I stammered out my story—arrived yesterday, incredibly busy, kids and RVs, backpack site, ranger said we could.

—Campers are required to register and remit payment *before* occupying a site, she said.

Well, I said, we didn't. But the ranger said. . . .

—I don't know who said that, it's completely against park regulations.

She had long red polished fingernails, and I noticed an emery board poking up from the breast pocket of her olive-drab shirt. I thought that was weird, for someone who worked in a *faux*-rustic hut in the woods.

—We sent out last week's receipts this morning, she said.

What was I supposed to say? I said:

—I see.

Then she didn't say anything. I said:

—How about if I pay for tonight and tomorrow night. We're leaving in the morning.

—That doesn't account for last night, she said.

—Let's pretend we weren't there.

But this was no laughing matter. We had as much as flat out stolen from the park, the department of natural resources, the state government, the taxpayers themselves. This stalemate might have gone on endlessly, except that the ranger woman seemed to get bored.

—I'll write you up for tonight and put the rest in petty cash, she said.

She began to write, and I said:

—I want some firewood too.

At the mention of firewood she assumed a cheery customer service air and directed me to a shed outside the reception hut.

—It's mostly seasoned hardwood, though there is some pine in it, she said.

By the time I left my head was spinning. I drove up the shore away from the campsite. I pulled in at a big resort with a neon fish and a flashing martini glass on its sign. It had a large restaurant and a bar that looked out through glass walls on the lake. There were a few people in the restaurant, coming and going from a huge salad bar. The real bar was empty. I sat at a table by the window and I drank, I don't know, three beers? two shots of whiskey? and ate stale peanuts and pretzels from plastic bowls. I could feel liquid seeping from the broken sunburn blisters on my shoulders, dampening my shirt. The pain was a steady buzz, but it was dull now. I charged the tab and then headed back for the campground as the light began to fail.

I missed the turn again. I was not driving so well. I walked partway to the campsite before remembering the firewood, and went back for it. I stumbled over roots in the heavy dusk. The path from the old highway to the campsite seemed much longer than before, but not long enough. I advanced toward the sound of yelps and cries. I saw flames through the trees.

The three were sitting around a fire with various forms of alcohol at their dispatch—the second bottle of wine, almost gone, cans of beer, the whiskey bottle, also near empty. They were practicing animal cries or something.

—What does the bear say? Frank said.

And K and Maria said:

—*Rowwrrr!*

They laughed. They looked strange in the firelight. They all turned when I tripped on a stone and flung the bundle of firewood ahead into the site. I saw that they had painted their faces. They had slashes and stripes, stars, circles, triangles.

—Just in time, Frank said. We're out of wood.

—It's a ceremonial bonfire, K said.

—We're going to go kill a bear, said Maria.

—We're warriors, K said.

They looked like frightening clowns. They really did scare me.

—We're going to kill a big old grizzly bear, and eat its heart and liver raw, Frank said.

—Heart and river law, Maria said.

K reached over and slapped Frank's shoulder.

—There aren't any *grizzly* bears here, she said.

—Are too, said Frank.

—Are *not.*

—Are too.

—Are not. What do you know, city boy.

—Oh, well, Frank said. Then we'll kill a sweet cuddly bear cub and put its head on a pole.

I had worried that they would see I'd been drinking, but

that wasn't a problem. I sat down on the log next to Maria. Frank and K sat on another across the fire.

—Let us paint your face, B, K said.

—War paint, brings down the power of the gods, Frank said.

—I think I'd rather not, I said. Will that stuff come off?

—Oh come on, Maria said.

—Don't be a pooper, Brycie, said Frank.

—Frank, I said.

—Only painted warriors are allowed at the bonfire, K said. Or are you a lover, not a fighter?

—Do we have a pole to put the bear head on? said Frank.

—I can fight, I said.

—You're the big medicine man, silent and wise, K said.

—He is, my friend is wise, wise beyond his tender years, and he's my best friend, blood brother. Let's all share blood, let's all be blood brothers.

—I'm nobody's *brother,* K said.

—Sisters, then, blood brothers and sisters.

—When are we going to kill something? said Maria.

—One big happy family.

—I'll paint myself, I said.

They brought me the cups and cans and a compact mirror. The pigment was weak, it just made a faint wash. I put black on my forehead, black under my eyes, bars of red and blue on my cheeks. I handed back the mirror.

—Is that all? K said.

She had thunderbolts on her cheeks and stars on her forehead and a red crescent on her chin.

—That's enough, I said.

—Well, that's better, at least.

—But it won't bring much of the power of the gods.

—If you steal the power of the gods, you die, I said.

—Not if you do it right.

—You die anyway. We're all going to die.

—So true.

—The power of Boscovitch, Frank said.

—Who's Bosco—? What's his name?

—We cannot speak of Boscovitch. Did you see Boscovitch?

—If you see Boscovitch you die, I said. Boscovitch is dead. He went to kill a bear and the bear killed him.

—Boscovitch is dead, long live Boscovitch, Frank said, lifting the whiskey bottle.

—What are you guys talking about? Who's Boscovitch?

—Boscovitch is no one. He's the ranger, maybe. He's dead.

—Let's go kill Boscovitch, Maria said.

—Boscovitch is already dead, honey, said K. Please pay attention.

The fire, unattended, had burned down low. Now Frank began adding wood from the bundle I'd brought, and K and Maria helped. They put in so much wood they smothered it. They took away wood, singeing their fingers, flinging smoking logs around the campsite. When the flames came up again they kept adding logs until you couldn't sit by it. It really was a roaring bonfire and they started doing their war whoops and animal cries and dancing around it. I rolled the sitting log back and sat out of their way. They tried to get me to join in, but I wouldn't. I drank one beer and that was all. I was already feeling a hangover headache from my fast drinks at the bar. They were still dancing but beginning to tire and I went by myself to the beach and looked at the stars. I saw one ship far out near the horizon. And tonight it seemed that I could make out lights on the Wisconsin shore; so perhaps the mind of God then too was comprehendible. I went to the water and splashed my face and drank from my hands. I drank as much as I could hold and wondered if it would make me sick. But it was so clear and cold and tasted like spring water.

Frank and K and Maria came down to the beach. They decided they would go in for a swim. I tried to tell them they were too drunk, the water was too cold, the bottom too slippery and rocky. But I felt no authority in my words and I couldn't argue long. They stripped under the starlight and a half moon just rising. I saw K's small pale body and remembered the cool skin warming and the pattern of freckles, the soft bob of her breasts, her chest chafed red, the pale orange hair between her legs. I wanted her and I supposed I could have her and I felt frightened and ashamed.

I should have stayed and seen them safely back from the water. I returned to the campsite. The fire had burned down to rich red coals that flared up yellow and blue. Voices came to me from the water. I climbed into the tent and undressed and zipped myself into my bag and only then, in the warm wrap of down, realized how cold I was. I lay with my head on a pillow of clothes and kept listening for them, tried to hear each one's voice, because two of them might not notice if one of them drowned. Then when I heard them on the beach I relaxed. I imagined what if K came into the tent and said did I want to, then I would say no, I can't, I'm too drunk, I'm sick. They came back from the beach and I heard them talking quietly, they were sitting by the coals. And I was drifting off and I jerked awake at the sound of the tent unzipping, and then I heard Frank's voice:

—Sweet dreams, he said.

He pulled out his bag and pad and closed up the tent. He hadn't taken his pillow so I pulled it over to me and laid my head on it and slept.

. . .

Dreams: The room it was the cave. There were women there and I was naked but I wore a shirt. I tried to have sex with one and then another and I couldn't, things wouldn't fit. They didn't laugh but they seemed impatient and they seemed bored. Then by the lake. It was big but it was like a big swimming pool, the shore was rough cement, the water tepid, dirty, smelling of chlorine. Little dead fish floating in surface scum, a gray film. I wanted to go in but the water was so awful. There was a woman and she ridiculed me for wanting to go in the water. I just dipped my toe in. I touched a dead fish.

I wanted to dream of black-draped nuns way out in the big lake, drowning or sailing with their wide habits spread, toes skimming the waves. Or Boscovitch striding on the waters, beckoning me to him. I wanted to dream of crossing the waters to join Boscovitch, for he called me, and maybe he would give me a ranger's hat of my own, and I would wear it proudly, and I would know the ways of the woods, and the deep water, and I would know the heart of the great father who strode a sky like a clear broad sidewalk.

I woke early and as soon as I woke I began packing to leave. Gray light filtered blue through the tent. I stuffed my bag and rolled my pad and gathered up the stuff strewn about as if I'd been living here for weeks. It was Monday morning. The picnic at Gran's had been Friday. I couldn't get time straight, though, nothing would fit, I couldn't remember what I had to do today, or this week, or ever. I opened the tent and threw everything out, climbed out and took down the tent and rolled and stuffed it before I even looked around. Three down sleeping bags on the ground, you might have said like cocoons but piles of colorful excrement was more in my mood. The campsite was a mess, cans and bottles everywhere and even the dirt looked disheveled. No one had bothered to hoist the cooler, but last night's rites had likely cleared the wildlife for miles. I identified Frank by his hair poking out the top of his bag, and woke him. He saw me and smiled with his mustache askew. I supposed he woke happy most days.

—Come on, I whispered. We have to get going.

He unzipped his bag and started to climb out. His face and neck were mottled from the face paint, most of which had washed away in the lake. I remembered that I had painted my face, too, and I found some water in a pot, scrubbed my face and rubbed it dry with someone's towel.

Frank wanted one last swim and I said fine. While he went to the lake I packed his things, gathered cans and bottles and garbage in a bag, arranged our things for the short walk out. The women still slept. K's red hair showed out the top of a green bag, Maria's black from blue. It seemed cowardly and wrong to leave without saying anything to them, but that was what I intended to do. It would leave everyone feeling bad that we had ever met, or maybe that was just me. K snorted and rolled over in her bag.

Frank returned naked from the lake, rubbing himself down with a towel. I showed him his clothes and he got dressed. Now all that remained in the campsite were two backpacks and two women in sleeping bags. I bent to pick up a piece of paper from the ground. It was the campsite registration slip. It had my name on it. I shoved it in my pocket. I reached for one handle of the cooler on which the box of camping stuff sat.

—Ready? I said.

Frank nodded, still sleepy. He took up his part of the load and we started down the trail.

The blue Bel Air, parked in the brush, out of sight of the highway, looked like a getaway car, and was. We'd got away to here and had to get away again. Or I had, did. Frank was an innocent bystander, just along for the ride.

There was fog again, not so heavy as the first morning. The highway was empty. I looked at my watch for the first time—it was still before seven. Frank went to sleep on the back seat. With no traffic we were in the port city in less than half an hour.

And then on up the height of land, the fog thickening as we climbed. Again the feeling as of crossing a border; but this time I was fleeing a beloved homeland from which I would forever live in exile. And it was my fault, nobody's fault but my own. Betrayal and treason, a plot that went so, so wrong. And begun all innocently, an innocence so deep it ran to craven cunning. And now I would have to suffer. I did not want to suffer. I repeat: I did not want to suffer. I felt no need to be ennobled that way.

At the top of the ridge we burst into sun and sky blue and broken cold clouds, a change in the weather coming, a front moving south. The lake and the port were left behind in fog. I left behind what was behind and saw only what was ahead, looked ahead down the highway as if the city skyline loomed there in a lens of quaking air; only a mirage of my mind. But what was behind rolled and boiled and chased me, clipping the wheels of the blue Bel Air, harrying me down the asphalt. What was ahead and what was behind would have to come together; they always did.

This and Not That

The solipsism of despair kept her mindless of danger when she began her night walks. She could fear nothing more than her shrinking room with its sinister angles, the dark insinuations of the rafters through the paint. In the movies it seems to mock common sense when the heroine flees from seeming refuge and runs to the dangerous dark. But the slasher's threat is nothing to the terrors within.

The problem with the room was, it was going to collapse into nothingness, and take Sylvia with it, so that where before there had been Sylvia in her room, now there would be no Sylvia and no room. It was going to crush inward from all its many walls and juts, and compress itself and Sylvia into an infinitessimal speck, and heedless of the scientific necessity of conservation of matter, it was going to disappear entirely. The room was bent on its own annihilation, and hers.

For this she paid $130 a month.

The first time she went walking this way was the night when the fraternity house shunned her, told her: too late, turn away. No hot apple cider for you this night (it's spiced and laced with rum), no popcorn by the fire, no ruddy-cheeked, orthodontic smiles to warm you. She felt the authority in that stony voice—it was Greek, after all, truly the voice of the ages. So she walked, followed the bridges east to west, west to east. She was aware of little or nothing around her. The walking brought no pleasure or ease. But when she returned to her room, hours after setting out, and fell into bed, then at least she slept.

She had no intention of doing it again, but something was changed by her going out that night, something was subtly begun. The next time she felt the panic building, the panic of being no one, nowhere, and she tried to forestall it by losing herself in those pictures she had tacked on her walls, she found that it did not work. She had seen that there was a place that was not this room and was not the imaginary city in the pictures. She wanted to forget it; she couldn't. She wanted nothing to do with the world beyond her walls.

But when the world within those walls informed her that it meant to crush her, then she went out. She went to the river again, only because that was where she had gone the first time, an option easily seized. She walked the bridges and she prayed, almost, for direction, a hint at purpose. She imagined she was being watched from high above. This exile was to fulfill some penance. So she couldn't go back before a certain time. When it was time to return, she would know. She didn't think at all about danger. It was late, it was winter, there was hardly anyone on the streets. She didn't care at all about physical hurt, might almost have welcomed it, to give form to her formless demons. Maybe the solipsism of despair actually protected her: if the world did not exist to her, then maybe she didn't exist to the world.

It required less and less to send her out as the months went by. If she couldn't sleep she got up and walked. If she had been studying long and was restless, she went out and walked. And eventually she even did it for pleasure. That she could recognize pleasure, want it and act to gain it, this was remarkable.

Then when she no longer felt she was sentenced to this, but chose of her will to walk, she was considering other choices, as well. There were six bridges she could cross to wind up back on the east bank. As she crossed them she played a game of choosing. She would be a famous scientist, supplant the big boss who sat atop the mysterious pyramid of research. She would become a doctor and minister to all in sickness and need. Or she would turn away from masculine, number-crunching, pill-popping medicine and become a midwife—she's pulling out babies with both hands, and with her nimble scrubbed toes; she's awash in a sea of fecundity, she's the fountain of life itself. She's a housewife and mother of myriad brilliant, lovely daughters, in a gingham frock. She's a poet, an alchemist of language, alone in a room of sympathetic light, with a bare wood desk clean and glowing, and she never erases a word.

She crosses from here to there. A choice, a denial. This and not that.

What do I want? she asks herself. She even says it aloud. And she wonders when it became necessary to ask herself that. The sure destiny that once led her has left her. Once she imagined bursting from a cocoon into brilliant butterfly wing.

Metamorphosis now will occur through a slow and painful unwinding of the thread, and who knows what's inside. As she walks she is weaving the fine gray silk back and forth across the river. It's like a loom, she thinks. It's her life loom. Slowly she is weaving up the fabric of her chances and her choices. Then she still has to cut something out of it to wear.

Walking became her therapy when she stopped seeing the counselor. She still thought of psychology with disdain, the more so since her mother went back to school. There is no less likely source of truth, she thinks, than those who offer it for a fee. These highly paid, certified professionals. But the counselor had helped, she had to admit. Seeing him had at least helped her remain steady. Who knows but without him she might have quite sunk.

The walking connects her to her city. It makes her believe, bit by bit, that she is somewhere. This somewhere is different from the pretty city in the pictures she once tacked to her walls. She will not say it is more real; it is equally real, utterly other.

One night without thinking she left the river and walked into the downtown business district. She found herself among office buildings and department stores, familiar sights, but so strange in this nightworld context. The mannequins pierced her with jaded looks. And she gave it right back. She passed her father's building, and the workers on the bas-relief over the entry seemed to be toiling in some Protestant hell where no matter how good your works or how great your disinterested industry, you would never be chosen.

Then on other nights she wandered among warehouses, down the old cobbled streets where legends painted on weathered walls offered *Better Health Through Better Plumbing,* and sought to comfort with *Don't Worry, Call MacMurray—Major Appliance Repairs.*

Burger cartons and aluminum cans chimed down the cobbles in a gray spring wind, scraping out a thin grim music. She felt like the discoverer of a lost and ancient land.

She became sensible of danger this way: As she walked down an alley, dark between blank brick walls, a door flew open, loud voices burst, just behind her. She gasped aloud, and jumped, and spun around. It was four men leaving from the back of a bar. Nicely dressed, trimly coiffed. Four gay men leaving an afterhours club. They saw her, and saw that they had frightened her. They comforted her. They all laughed, Sylvia too. They said she'd scared them, as well. They told her she shouldn't be out alone so late, especially in a place like this, and they walked her to a lighted main street. They offered her a ride, they insisted, but she assured them she didn't have far to go, and she left them as they continued to insist.

So now when she walks she is aware of danger. She knows that anyone would say she is crazy. But it's just because she's no longer crazy that she is aware, and scared. She isn't always scared, but she is wary, and less likely to become lost in her thoughts. That is a loss. She understands how it is that she could walk all those nights without fear, but she still marvels at it. She thinks of how she used to climb trees when she was small, way up to the topmost swaying branches, almost to the sky. She never knew or heard of a kid who was badly hurt falling from a tree. She never knew anyone who fell. Then maybe the lack of fear did protect.

Now she's a little addicted to the danger. When she started, when she was writing in her journal such things as *Each day the bed of gray ice draws me downward,* she dared herself to be on the bridges, to see that cold bed. Now she walks to face a different dare. Or maybe, she considers, just to be contrary.

The day she met Carla at the restaurant, and Carla invited her to her friend's party, she walked that night to choose. This was a weighty, troubling choice, more grave than scientist, doctor, midwife, poet. Who was this Carla? Someone she hardly remembered, a part of her past, with which she was not on friendly terms. Sylvia might disappoint; she wasn't Zelda anymore, she wasn't brilliant, she wasn't pretty. All that in the past. It might be an invitation to embarrassment and regret.

She dared herself to go, and went.

She went to the party. She came home queasy and triumphant. She went out to walk off the queasiness, and to celebrate. All the people, who knew each other, whom she did not know. Deserted by her friend, she had wanted to bolt, but she stayed. A great victory.

She had met an odd man. He was by turns charming and graceless. He baited her, and she stood up to it. He went away, but he kept coming back. If there were signals there, Sylvia didn't know quite how to read them. But when she walked that night the game of choosing she played was this: He'll love me, he won't; I'll love him, I won't. She knew she might never see him again.

Fourth of July and her room so hot. The family picnic an unredeemed disaster. She's angry with everyone, mostly herself. She doesn't know why she acted that way, why anyone acted the way they did. He said he would come here, but she doesn't want to see him. She can't stay here, she thinks she'll go out and walk. But this is something she won't walk off. She doesn't want to think about it, not tonight. There's somewhere else she can go, someone else she can be with. She'll go to see Carla, and have a drink. Watch the fireworks, maybe.

Him or her. Here or not here. Him or her or herself alone. Do or not do. Be or not be. Wondrous, and grave, to choose.

Part Four

Cook

Sylvia, it was because I waited for you the whole long lifetime of that firefly fireworks night, and writhed in dreams of the passionate making-up I had envisaged, and you never came, and you were with another; because I wanted to carry you away, and lay you down, and take you in the weeds, to show you what truly mattered, to make it reach you, but I did not.

Because I wanted to fell the tree of good and evil, but the squirrels beat me to it.

Who was it that I took, by whom was I taken, on that rock like the dome of a great forehead? Almost you and almost me atop the deep genius of stone. A piece of the secret desire: does it require only an action, and look indifferently on its object?

. . .

I walked across the loading dock and in through the restaurant's service entry. A young guy with long dirty hair, wearing splattered whites, came around the corner as I moved between tall stainless-steel shelves full of pans and plates. He stopped and prepared to confront me, then Carla appeared behind him and said,

—Hi Bryce.

And to the young guy:

—He's okay.

In a joking tone, but the guy didn't say anything, just moved past me, removing his apron and jacket, threw them in a bin of dirty linens, and went out where I'd come in.

Carla came toward me and we kissed lightly on the lips, she holding her wet hands out as she raised her face to me. Her face was flushed and damp from the steamy heat of the kitchen. Her hair was pulled back in a bushy ponytail, and she wore a red-and-yellow cycling cap as a chic hair restraint.

—We're almost done here. Wait in the dining room. Grab a beer on your way, in the ice there, she said, pointing a wet finger.

The dining room was empty, and all the tables cleared and stripped. Except for one, a window table, where a couple sat drinking the last of a bottle of wine. They were talking and laughing and looking intently at each other, and holding left hands while their right hands raised wine glasses. The woman was pretty in a waifish way, with short black hair under a black beret, a black sweater, slender white neck and deep red lips. The man was in black too, the uncolor of the avant-garde. His sandy hair was longish on top and shaved on the sides. His face was square and quite handsome and he wore fifties-style horn-rimmed glasses. He would tilt his head in a boyish, demurring way at something the woman said. She would lunge forward with an utterance and he would move not back but away slightly, a movement like a fencing feint. Their eyes were bright, they were well-fed and wined. They had paid their bill and the tip—quite generous, it looked—lay weighted by the ashtray. Their legs crossed under the table, their ankles touched and rubbed. She wore black hose and black slippers; his dark cuffed trousers were hitched up and showed a little pale calf, then black socks with red diamonds, slumping, and pointed black wingtips, a little scuffed. I found those ankles deeply touching. A light patch on her nylons where her ankle bone stuck out.

It was well past closing time, nine o'clock on this Sunday night. Another young sullen guy with blond hair in a long ponytail finished mopping the kitchen and leaned in the kitchen entry, arms crossed around the mop handle. He glared out into the dining room, and killed the happy couple with every look. Whoever was still cleaning up the kitchen was making an inordinate lot of noise, another signal to the lingerers, I guessed.

Shortly the couple's wine was gone, and they pushed back their chairs and swung out through the glass door to the empty sidewalk. Their arms came around each other, they turned naturally into a kiss, and gracefully parted, and went on their way. It was exactly like watching a scene in a movie. The mop man had followed the couple to the door and locked it, then he tipped chairs on tables and flailed the mop along the floor like a rabid wet animal while I kept conscientiously out of his way.

When he was gone I tipped down two chairs from a table in the center of the room, and sat and drank my beer. The noise in the kitchen dwindled, the back door opened and slammed twice, and all was quiet except for Carla talking to herself in a low voice at the back of the kitchen.

Now it was the weekend after the Fourth. I hadn't seen Sylvia yet. We had talked on the phone. She said she was sorry and I said I was sorry, and neither said what for. I said I was busy with work, that I'd been neglecting things, and she said, me too.

The morning I returned to work I learned that the neighborhood had been officially declared blighted. I walked the neighborhood streets seeing everything through the new frame of blight. In some way the neighborhood sordidness was a solace, in that it seemed to mitigate or at least explain my own crimes. Our dwelling place had been called blighted, and our homes and our children a pox on the earth, our very flesh an abomination. So what do you expect. I went to gather the weekly police report. The liaison sergeant's criminal rosary became especially virulent now that it arose from blight: burglary, robbery, theft, assault; criminal sexual conduct in the third, second, first degree; prostitution, solicitation, domestic violence, disorderly house. Our house, as our lands, dear Lord, is a shambles.

And in the clanging, disjoined heap of my emotions, the neat inverted pyramid of the news story was a rock of certainty. Whatever I wrote was in some way so. What came first was most important, and what dribbled out in the final 'graphs could be deleted, made not to exist, without altering at all the stony factual state of things. I craved this certainty of artifice, because what was real, if such a thing existed, seemed beyond reach or understanding. In a well-crafted piece of factual journalism all the essentials could be distilled to a lead of thirty-five words or less. I knew it was fake, but it was at least informative. The readers weren't concerned with my epistemological qualms. And I suppose that deep down I wasn't, either; epistemological qualms were another veil. The mind is like an airplane that can doodle in an endless holding pattern and forget entirely that the point is to land.

Carla came out of the kitchen still flushed and flustered. She carried a cosmetics bag and clean clothes. She motioned toward the women's room as she walked that way.

—Just give me a minute, quick Polish shower, then I'll cook us some dinner.

I said fine, don't hurry, and she stopped with the restroom door pushed half open and said:

—Oh, you're not Polish, are you?

—No, I said.

—No, of course not. I'm Polish, sort of. Polish Jewish, Ukrainian, German.

—I know, I said. Just get cleaned up, would you?

—Right, she said. You could set the table, if you can find the silver.

Her wave merely indicated the kitchen, and the door swung closed.

I found the silverware, and a tablecloth, and linen napkins. As I set the table my mind ran back to blight, specifically to one tangential victim of the blight, one Mrs Ella Oie, or more properly Mrs Hubert H. Oie, a widow, age seventy-seven, a Red Earth resident of forty-six years. Mrs Oie called the newspaper outraged, often. She called the police, she called her city council member. I went to see her this time, and to the incongruous accompaniment of silver spoons tinkling translucent china teacups, the shy crunch of shortbread whose crumbs fell more modestly still on an only slightly yellowed lace tablecloth, she told me how she had watched a whore give some suburbanite a blow job in a car parked outside her window (. . . *the alleged black female perpetrated an apparent act of oral sodomy on the purported white male in a seeming late model Chrysler supposedly blue putatively parked on the suppositional street*—in the language of the police report, roughly). Ordinarily we didn't bother much with the all-seeing Mrs Oie's complaints, but there was a big prostitution sweep on, and she was our representative victim of the victimless crime. I wanted to say, pull your blinds, Mrs Oie; I wanted to ask, why do you stand darkling at the midnight window, Mrs Oie, to witness these pathetic acts of sex and commerce—is that insomnia back? But we won't be made prisoners in our own homes, and we should be able to look out our windows at our sovereign parking spaces without being slapped in the face by acts of sordidness, unless, of course, we want to. But isn't there room in our great democracy for the sordid as well as the righteous? Isn't that our last best hope?

No, I sympathized with Mrs Oie, I liked her. And she thought I was a nice young man, she told me so. I sipped her tea and ate her shortbread and didn't mention that my cup was chipped. I sympathized with her teacups, her silver, her lace. That was the real story of Mrs Oie. No: The real story, specifically, was how one crumb of shortbread had clung to Mrs Oie's lip, her thin lower lip marbled pink and gray with decrepitating lipstick like an ancient weathered fresco, teetered on the lip and then tumbled, glanced off the handle of the sterling teaspoon resting on her saucer, hopped onto the tablecloth and rolled to rest on an old faint tea stain the shape of Ohio in a band of sunlight which swelled and ebbed as the slight breeze lifted and dropped the blind drawn against the afternoon sun, and there glowed golden, the crumb, glowed rich and golden, on what would have been the shore of Lake Erie, at the site of what is now Toledo. But the crumb was a crumb in the wilderness. I observed this as Mrs Oie described I recall not what. I observed this and I knew that this was the real story of Mrs

Oie, and I knew I could never get all that into a thirty-five-word lead. I might be fired if I tried.

This restaurant silverware was not nearly so nice as Mrs Oie's. Still I tried to set an attractive table. I found a candle—wax in a glass, really, with enough black wick—and lit it and put in on the table. Carla came out looking scrubbed, in clean clothes, a long red blouse untucked over pegged black jeans. She moved into the kitchen and put on a clean apron, and said,

—How about a fish? Could you pour me a glass of wine—white. There's an open bottle in the fridge behind the bar. Just bring the bottle over, I'll need it. For the fish, and otherwise.

I brought the wine while Carla arranged pans on the big black range. Then she disappeared into the walk-in fridge and returned with a glistening rainbow trout, butter, shallots, ginger, pea pods. She examined the fish briefly and said,

—Clear eye, red gills, that's how you tell it's fresh.

With a big chef's knife she reduced the ginger and shallots to slivers in seconds, the knife going *tocktocktocktocktock* on the cutting board.

—And grab an ashtray, will you? And do you have a cigarette?

I lit the cigarette for her as she floured and seasoned the fish and browned it in a pan with olive oil and butter. I put the cigarette in her mouth as she lightly prodded the fish. She laughed, and said,

—Do you remember in *The Thirty-Nine Steps,* when Robert Donat brings the mysterious woman home from the music hall, and first thing they go into the kitchen and he cooks dinner for her? He throws a great whole fish in the pan, a flounder, I think, and nothing else. He's smoking a cigarette while he cooks, he slings the fish on a plate, he slices a thick hunk from a loaf of bread—and that's a meal. And he hasn't even finished the cigarette. I always loved that. I thought, that's my idea of cooking.

—So when you're a famous chef and they come around asking who was your greatest influence you'll tell them Alfred Hitchcock.

—Right.

I hiked myself up on the counter beside the stove and I found myself growing calm as I watched Carla cook. While the fish browned she trimmed the pea pods. In an oblong pan she heated oil and laid in cooked noodles to brown into a cake. She removed the fish, threw in the ginger and shallots, and in a moment added a slosh of wine, swirling the pan and then dipping the edge toward the burner so the wine ignited and a great pale flame, yellow, blue, and red flashed up, a dreamy fire that lasted only a second. A gelatinous glob of broth went in and melted to iridescence, a toss of salt, a grind of pepper, butter, then back went the fish to swim in this ambrosial soup. She covered the fish, flipped the noodle cake, sautéed the pea pods, tossing them in the pan with a nimble motion of the wrist, finished them with a little salt, a little sugar, a little vinegar, and a reddish-brown spice with a wild, citrus-y smell that she said came from a thorny bush in the Sichuan province of China—the Chinese called it flower pepper, she said, and I could see the gnarled fingers of old stooped peasants winding in among blossoms and thorns to pinch out the berries, and I could see them drying in a hazy Asian sun.

The trout came out of its sauna. The noodle cake went on an oval platter and was bathed in sauce, and the fish was exalted on top of that, and the pea pods made its laurel wreath. I was hungry. Carla smiled at the fish and at me.

—Let's eat, she said.

—Let's, I said.

She brought the platter out to the table, and I brought the wine and a baguette and butter. I had set out two plates, but Carla moved them away.

—It's too pretty to portion out, she said. Do you mind if we just share it?

With the platter in the center of the table we leaned toward each other and reached in with our forks to pry a bite of fish flesh from the backbone. The trout tasted sweet and clean and the rich sauce reduction set it off to perfection.

—Good, I said. Really good. Wonderful.

Carla tipped her head and nodded, and her expression was critical but approving. She raised her wine glass and I raised mine. She said:

—To the fish.

—To the cook, I said.

The lovely construction began to collapse as we pulled away shreds of noodle from the trout's pedestal. Carla said:

—I love to eat whole things. Whole fish, oysters, clams, crabs, quail, artichokes. Whenever we change the menu I try to put whole things on. I don't think anyone has noticed. I feel like I'm writing a secret book. Supposedly there's a Chinese dish called "Three Squeaks"—it's tiny baby mice, live. You pick one up with your chopsticks—one squeak; you dip it in sauce—two squeaks; pop it in your mouth—three. Eating wholeness, eating yin-yang. You are what you eat, after all.

—I didn't know you had a philosophy of food.

—I said it was a secret book. If you cook and you merely pay attention, you get a philosophy. All the wonderful things of the world come into the kitchen, and you shape them, in so many ways. At first, out of college, I couldn't reconcile myself to throwing a degree away in food service. Eventually I figured it out—I'd be throwing myself away if I did something else for the status or appearance of it.

It had been a long time since I had talked to Carla, just the two of us. The last time, I realized, had been at the party where I'd met Sylvia. I had forgotten how much I liked her. We caught up with each other, though nothing much momentous was admitted. I said I had been up north over the Fourth. Carla loved the big lake, she was jealous that Frank and I had gone, and why hadn't we called her. She asked, laughing, about the car. I told her about the picnic with Sylvia's family, and my epic struggle with the chaos of Gran's yard. I enjoyed telling it. Carla thought the entire car episode, from beginning to end, was hysterical. But of course it wasn't ended.

—That car is your destiny, she said. You'll drive it to fate's end.

She said she had seen Sylvia on the Fourth, but Sylvia had hardly mentioned the picnic. They had just had a quiet evening together and watched the fireworks. Sylvia seemed calm, Carla said.

It was true what Carla said about the car, but it wasn't the car she meant it, it was Sylvia. Unassumingly, Sylvia had become the center of both our lives. Sylvia was why we were sharing this fish. Sylvia was the fish. We'd eaten all the flesh on one side, and we flipped it over to get at the other. This side wasn't so pretty—much of the brown skin had stuck to the pan and pulled away. Just the same, it was delicious.

—So you two watched the fireworks, I said. Did you drink gin & tonics?

—Gin & tonics? I don't remember. No, I think we just had beer, maybe some wine. Why?

I told her what I remembered about the end of the world, and she laughed. I asked where she had gotten the idea about pitchers of gin & tonics. She said she didn't know, it was just a pretty idea.

—Anyway, it wasn't the end of the world. At the end of the world I guess you can drink whatever you want. Gin & tonics *and* martinis, and pour your wine in there too.

The fish had been stripped of greenery—the pea pods were crisp and sweet, with a depth of flavor from the arcane vinegar Carla had used, and the Chinese pepper had a brilliant, slightly numbing flavor. The fish itself was now just a skeleton askew on shreds of noodle. The way the fish's head was angled, with its mouth slightly open, with its flat milky eye, it looked rather downcast. But the way its tail was curled, upward, and glistening with sauce, gave an impression of fishy frolic. I picked at the bones and didn't look at Carla when I said:

—Are you in love? Are you—what was it?—smitten, or enthralled?

—Am I what?

—You know.

Carla touched her mouth with the white cloth napkin. She reached for the wine bottle and poured into both our glasses. She picked up my cigarettes and gave me one and took one for herself. She lit hers with my lighter and I lit mine from the candle.

—Yes, she said. All of the above.

—Uh-huh, I said. So?

—So? So what? I was in love with you before. Now that's all okay, isn't it?

—That was just because. . . . You weren't really in love with me. It was college. We were young.

Carla laughed.

—And now we're gray and ruined, teetering on the grave. I really was.

—But that's not the point.

—What is?

—What is. Is she in love with you?

—I don't think so. I don't know. I don't think she's inclined that way, at present.

—You think she could be.

—You know what I think. I think anyone *could* be.

We sipped our wine and smoked, and then I said:

—Now is when you should say, Why don't you ask her?

—What's the point in that?

—We're awfully short of points here.

—Just so, said Carla. Still, no point in forcing one. I don't think it has anything to do with me. I'm somewhat resigned. There's a boy or a girl out there for me. Though I have had some bad luck. I would say, Bryce, that we're still young. Though you, you were never quite entirely young. Maybe because of your dad.

—Maybe, not necessarily in the obvious ways. But that's not the point.

—There you go again.

—Oh, shit, I said, and I lofted my white napkin softly over my shoulder.

I was feeling the wine, and I wanted there to be a point. Still, I felt safe from harm here in the empty restaurant, at our table in the rectangle of light the bright kitchen threw out. So I started telling Carla everything that had really happened, what I thought had really happened. These things had happened one by one, and if I could sort them out perhaps they would tell me their point. It wouldn't be just a mass of impressions wrapped up in a dirty secret. I told her what had gone on between me and Sylvia at the picnic, as far as I could tell. I told her about the night of the fireworks, and I told her what had happened at the big lake. I tried not to analyze or attribute motive or psychologize or excuse. As if what had happened would speak for itself, as if all these events were neutral and equal in value, like identical white eggs in a row. Carla listened. A couple of times she asked for clarification. And when I finished she sat nodding. And then she said:

—Don't tell her.

I took that in. It was just three words. But words of weight.

—How can I not tell her?

—Why would you tell her?

—Because of, you know, honesty. Because I love her.

—I think Sylvia values loyalty above all else.

—Above honesty?

—Sure.

—But I've been neither loyal nor honest. Disloyalty is the cause of my dishonesty.

—It's not something you can really fix. Maybe in time, but . . . I assume you're hoping for a future in this?

—Yes, but, it all seems so unreal.

Carla nodded. She stood up and went to the bar.

—Coffee? she said. Brandy? Something else?

—Coffee. Brandy.

She turned the coffee maker on, and came back with a tray holding a bottle and two warmed snifters. She poured, and we swirled. I breathed the fumes, too deep, and coughed.

—I don't mean this unkindly, Bryce, but, about Sylvia, I don't think you have a clue.

And she expanded, saying that Sylvia was different, she'd been all alone since she was eighteen (and probably since long before), just trying to survive, she was too smart for her own good, probably an honest-to-god genius, in that she saw the world differently from anybody else, that she was a fatalist with the stoniest will to survive that Carla had ever seen, that she wasn't cut out for the girlfriend role, she was bewildered by love and romance even as some part of her, that old Zelda part, clung to the starriest notions of romantic fulfillment, that she was sweet and vicious and mystic and pragmatic, and what else.

—But, really, she's just a girl, a woman, and a girl, and, yeah, *human*. Maybe that's the important part.

Carla sat back, sipped her cognac, lit a cigarette. She was apparently satisfied with her rendering of Sylvia. I had to congratulate her. I had to agree. I knew all that, I truly did. I just didn't trust it. It sounds ridiculous, to be so skeptical of your lover's humanity. Unless humanity seemed to diminish her? How could it? Of course it could, but that was the hard part of the bargain.

—You act as if she's some exotic animal, Carla said. It reminds me of some old *Star Trek* episode, where the aliens want to love and shelter the humans, keep them as pets. And Captain Kirk says, no, no, we could never live in captivity, to be human is to be free! We'd rather die! It's pretty cliché, I know. You could probably dredge up some profound literary reference to show the same thing.

I suppose I could have, but I didn't try, because it seemed that cliché was precisely the point. At last, a point. I thought of secrets, deep and dirty. I believed that one's true secrets were never told. I thought of eggs in a row, the *WhoWhat-WhereWhen;* but *W* #5, the *Why* of it, the undissectable motive, if it was an egg, it wouldn't hatch. But still, I had a notion, a pebble dropped down a mountain and it would find its own way, a notion of why, and it was because I knew Sylvia was different, I had set out to hurt her to make her like everyone else. Not that I thought she didn't suffer. But I wanted her to suffer not in some spiritual, existential, mathematical way. I wanted her to hurt in such a way that she saw herself in pop songs. A brief fantasy of driving Sylvia in the blue Bel Air through the wide countryside while she weeps and wails with the AM radio. *Listen to the sad songs that the radio plays, have we come this far far far to find it so cliché. . . .*

—What? Carla said. Fa fa fa?

—Just an old song, I said. Not that old.

Carla went to pour coffee. I got up and joined her.

—Let's sit at the bar, I said.

It felt appropriate. Let's sit at the bar and talk of old times and sad lost love. Off the rebound from our disastrous and very brief fling—really just a bad patch in our friendship—Carla had become involved with a woman whom none of us—her usual circle of friends—ever really met. I remembered seeing the two of them from a distance, one spring day on the grass of the main quad. The woman's hair was cut short and uneven, you would have said punk, then, but she dressed like a farmer in baggy jeans and boots and flannel shirts. That day on the spring grass I watched as she and Carla argued, or rather as the woman apparently lectured and accused and Carla desperately defended. Their relationship was one crisis after another, and I think in the end it was Carla's lover who called it off. Carla wasn't committed enough, to anything, was the cause given, though the breakup hurt her terribly. It lasted a couple of months, which at any rate was a good deal longer than Carla and I had survived as a couple, which was about two days. But something happens and then everything is changed, not just the now and the will be, but the was, as well. So one was never really safe. Even the past was nothing solid. I contemplated inaction, a Buddhist mountaintop, gongs and prayer flags. Turn away from desire, let the wind wish for you. Wanting and doing led to tragedy and heartbreak. Tonight with Carla the mood was somber but assuring. The restaurant was a haven, the kitchen like a womb. But I would have to walk out the door, eventually, into the dire flux. That's life. It struck me ironic then, I guess, that it was here that all this had started. A woman poured cream into her coffee, and Carla's heedless imagination leapt to a high school chemistry lab, a girl babbling in German over molecular equations. Couldn't blame Carla, because Carla was true. For Carla the girl in the chem lab was a frozen symbol. But Carla didn't want to leave her that way, she wanted to know her now and as she was. For me—Carla was right—Sylvia was still that green-eyed girl, that prodigy. Didn't I want to love the Sylvia who dressed badly, whose skin flaked, who actually cared whether her little poem was printed in my dumb little newspaper? The fault was with me, then, I thought, that I feared the risk in mortal faulty love. That night of Memorial Day, when I had held her bare shoulders and declared my love and said I would not let her slip away, I was maybe talking to myself. And had it shocked or frightened her? Why say it hadn't, just because she is Sylvia.

Carla poured more cognac into our glasses. She could drink at least as well as I, but I saw her growing melancholy.

—Then that night, the Fourth, I said. Did Sylvia stay at your place? I waited most of the night for her. I was worried.

A lie!

—Yeah, she stayed with me. We slept on my futon, both of us. Girls do that, you know. When she was asleep I brushed her hair back so I could see her face. That's all.

—Well, I wasn't really worried, I guessed she was with you. I was even sort of glad she was with you, if that's where she was, which she was.

—This isn't as hard as you're making out.

—Maybe it wasn't, but now it is. I think it's even worse, because right now I feel okay. But soon, I won't.

—Oh, you'll survive. You may even be happy again, someday.

We laughed. Carla said:

—But why did you have to— Men and their dicks, it's where all the trouble starts.

—Women will always blame it on dicks, just because dicks are so obvious. And because women don't have dicks, you're automatically excused.

—I suppose that's so.

—No, I said. It is dicks. In a sort of symbolic way.

—Symbolic, bullshit.

We talked for a while about dicks and symbolic bullshit. Then we cleared the bar and the table, and because it was a lovely cool night I walked Carla to her building a few blocks away. Outside her door we kissed and hugged long, as we hadn't done for years, maybe. With my face in her hair I smelled cigarettes and ginger and a freshness that the air hadn't held all summer. I felt her holding me, and I thought, she isn't a genius or a creature of mystery, but— And I wondered. Then we parted and kissed again quickly.

—Nightcap? Carla said. What time is it?

We both looked at our watches and Carla said,

—Oh.

—Late, I said. I'd better go.

Walking back to the car I thought of the couple in the restaurant earlier that night, and I wondered where they were now and what they were doing. Had they gone on to a dark intimate bar, maybe one in a fine but unfashionable hotel, and mellowed to where the woman stopped lunging and the man stopped feinting, then walked slowly to—whose apartment?—her apartment, just a block and a half, then turn left and another half-block, thinking how lovely it was to live so near the fine but unfashionable hotel, and up the swept steps and through the glass doors and winding up the grand staircase past quiet dark doors, their steps soft on the elegant worn carpets, and keyed the apartment door and heard the tumblers clack and roll with a sound as solid as Lloyd's of London, and into the rooms of dark carpets and white draperies, and fallen together in warm fluid love? Their stockings, her black nylons, slightly pilled and with a run on the thigh where it didn't show, and his wooly black socks with red diamonds, lay mingled together on the floor. From a distance you couldn't tell they were stockings, but the black tangled clump, the glowing red rhombuses, were the very essence of intimacy. There is a balcony off the bedroom, but the French doors are closed. They may be opened afterwards to let in the night air, the scent of summer and dark city. What would Mrs Oie think, if she lived across the street? Such things could happen. But so could other things. It may have been that they parted not long after leaving the restaurant, saying it's late, I've drunk too much, tomorrow's Monday. And alone in their separate rooms the magic balloon of love, sex, and joy had collapsed, dropping each back into his, her, own sole self. He in his yawned and put on the TV and fell asleep in the chair. She in hers scrubbed off the make-up, washed the waif down the drain, fell into bed thinking of Monday morning at the computer terminal.

Then here I was at the car. My fate, Carla said. I drove back toward blight through empty streets, with a self-consciousness peculiar to being alone on the road at half-morning in the middle of the city. I thought of Carla, the woman and her lover, and of Sylvia. Waiting at a long light without another car in sight, thinking I could run this red, but not really tempted to, it came to me, why I loved Sylvia: Sylvia knows that beauty dwells in sadness. Just that. Beauty dwells in sadness. At first it seemed tragic, then I came to think: It's still beauty, isn't it, no matter where it lives. No need to mope. It's beauty wherever it dwells, and what we live for.

The light then changed. And here was blight. Secrets are not told. Language is treasonous,they say, subverts our meaning and intent. Language is a sledge hammer in the jeweler's shop of the soul. But nice to think that the soul is a jeweler's shop.

If I told her I would tell her and not explain. How could I say the why, the wild guess, the dismal fiction. No way to convey the emotion in commission, the solitariness of it, the contextlessness like a desert around the heart. *It didn't have anything to do with you, darling.* The adulterer's hackneyed plea. But, you know, true. I could bring it into some context with Sylvia, but that would be a fiction, too. Of course it would mean something to her, even now, when she didn't even know it, it meant something to her. But it meant, would mean, different things to the two of us. I thought of Frank's notion of derailing time by considering every possibility. But here I was home, I was tired. I decided to let time continue, at least for the night.

River

Time continued then and in time brought me to Sylvia's door again. And though so much had happened, it was as if, as I climbed that famous fire escape, time in its reckless way had thrown a loop. It was the Fourth of July and my darling awaited me at the top of the steps and everything would be all right. It was a warm evening, sticky. She sat barefoot, in a thrift-store cotton sundress. She looked hot but as if she liked it. She was drinking a beer and smoking a cigarette. Her hair had grown to where it always looked mussed. There was a persona at work here: Summer Sylvia. I bent to kiss her and she held my neck and pulled me down to her, and she lay back and I descended to her between her spread knees and lay on top of her on the metal grate landing. We managed it as if Sylvia had been thinking of it for some time.

Ending it was, as usual, more awkward, but we got up and went inside.

And when we lay wet and spent Sylvia lifted her head and said:

—We missed us.

We did. We greeted us this way. This we knew, knew how to do, so easy and so sweet. Through the open door came a diffident breeze that dried the sweat on belly, chest, and thigh, and raised gooseflesh, a tentative shiver. Sylvia went to shower. I rolled over to let the breeze touch my back. With my face turned to the wall I thought of nothing.

Rolling over again, I looked through the door to the hazy tops of the big green trees. The leafy mass stirred oh so slightly. Here and there a naked, bleached branch protruded, bony fingers in the sky. It was late July, fullest summer, but already there were cool mornings that breathed autumnal air. In my neighborhood rounds I often passed a community garden, and had watched the level hope of tilled earth spring into green, watched beans race weeds, and now the plot was full to bursting, tomatoes pressing the fences, squash vines with dinner-plate leaves and imperial intentions threatening everything in sight, eggplants displaying absurd, exotic tumescence from strong purple branches; all like a vegetable clock, impending noon, morning long past and not to come again in this season. The beginning of the end; but beginning, doesn't it?, holds ending, ticking in its genes.

Sylvia returned. The opening door brought a freshening of the breeze, and its closing stopped it. Beside the bed Sylvia dropped her robe. She lowered herself onto my back. The touch of her cool breasts brought a shiver, then heat as she pressed herself to me. She kissed my neck and cool strands of wet hair brushed my shoulder. She kissed my neck, then my cheek, then she stood, and said:

—Ugh. You're sticky. Go take a shower. Then let's go somewhere.

I rolled onto my back. Sylvia, her back to me, stepped into panties, then lowered the light dress, green and blue, over her head. I sat up, reached for my clothes and Sylvia's damp towel. I wrapped the towel around me and carried my clothes, and at the door looked up and down the hallway, and walked quickly to the bathroom.

Sylvia was on the landing when I returned. I leaned in the door and said:

—What do you want to do?

—I don't know, she said. Are you hungry? We could get something to eat. But I want to be outside. Let's pick something up and go down to the river.

—Doesn't it smell? I said. The river usually smells this time of year.

—It won't be too bad. Let's try it. We can always leave, or just sit up on top.

We walked to the campus grocery and bought a roasted chicken, coleslaw, bread, grapes and apples, blue cheese, pickles. Sylvia was hungry. She picked the food and I paid for it. In the liquor store next door, in the wine cooler crammed with alcoholic soda pop, apple and strawberry wine, I found a bottle of cheap Italian white, a big bottle, with a screw top.

Then we drove to a place that Sylvia had in mind, on the west side of the river, south of the University. We parked on the river road and followed a trail down the steep bank, descending in dark golden shade. A humid evening was just coming on. The trail had steps, in parts, and the odd stretch of rusted handrail. So it had once been an official path, but now it was not maintained. Along the river here there were caves in the limestone cliffs, and sometimes people went in and didn't come out, so the city did not encourage traffic down here. Down by the river was where bodies appeared beneath the bridges. Often the bodies were hard to identify, because they were people that nobody knew. A man in a blue windbreaker, with a scar here or there. Hobo jungles down here, and riverine dens of iniquity to serve any taste, it was said. Toward the bottom the path was well blazed with cigarette packs and beer cans, condom wrappers, and the potent, unmistakable shape of used condoms, something horribly animate in the way they sprawled amid the leaves and trash, like once-live things now dead, and far from home—jellyfish drying in the desert. I remembered Alison's discovery in Gran's yard; nothing about these things in this place suggested love.

Nonetheless, once we were settled on a shelf of pinkish sandstone, with intrepid saplings clutched to the cliffside over our heads, we ate with good appetite. The river ran shallow below a lock and dam, and it was split by a low island covered with scrawny poplars and scrub. The island lay between us and the eight-foot channel that carried the river traffic. On our side the water was narrow and sluggish, and marshy grasses grew along the island's shore.

We pulled apart the chicken with our fingers. We had plastic forks for the coleslaw, but we'd neglected any other utensils, or napkins, so Sylvia's multifarious feast became a free-for-all, and our fingers dripped with pickle juice and apple juice, and became gluey with blue cheese and chicken fat. We shared the wine back and forth, tipping it up between our relatively clean palms, like babies with a great big bottle. We ate in a kind of measured frenzy, not hurrying, but rarely stopping, and hardly talking at all. It struck me that Sylvia and I had hardly spoken to each other in weeks. As we crumbled off bits of blue cheese, and dipped for pickles and dismembered the savory bird and tore off chunks of bread, we both knew it was time for talking. So this was perhaps the final feast of the condemned. But I didn't feel condemned. This meal was weird, but it was good, and fun. It was a lovely evening. The river didn't smell too bad. I looked out at the island, and saw a great blue heron at the upstream tip, petrified in the weeds and peering ferociously, its big beak tipped downward. And then a white egret appeared at the downstream end of the island, and looked upstream, and froze. These two tall birds stood as bookends on the island.

I should have felt worse about the terrible betrayal I hoarded, which I had not decided whether to tell or not. It was like something that had happened in another life, or something overheard, or read in a book.

Sylvia stood and brushed bread crumbs from her lap, and slid off the ledge and washed her hands at the river's edge, rubbing sand between her hands. The birds paid her no mind. She flapped her hands in the air, then patted them on her dress. She came back up, and I went down and washed my hands too.

As I climbed back onto the sandstone shelf Sylvia reached out a closed fist to me.

—Here, she said.

I opened my hand beneath hers, and she dropped into it a bit of yellowish stone. I turned it over in my palm.

—Trilobite, she said.

A dark, smooth lobe curved out of the rough limestone. The lobe glittered faintly. It looked dense and cold. She reached behind her into a pile of limestone debris that had gathered at the base of the cliff. She handed me a triangular piece, the size of a slice of pie.

—Bryozoa in there, and a crinoid bud, and who knows what.

I saw shapes like tiny scallops and clams, and something remarkably like a flower bud. Sylvia turned, and, kneeling, scrabbled at the bottom of the pile, picking up with thumb and forefinger things too small for me to see. Then she turned back and dropped into my hand the things she'd found. They looked like tiny stone cogwheels, intricate bits of a prehistoric machine.

—They're rings from a crinoid stalk, she said. But it was an animal, like a sea anemone. Ordovician, from the Precambrian sea. A warm salt sea.

—How old?

—These fossils? Five hundred million years. Down here at the bottom, a billion.

Rising up behind us, and across the river before us, one billion years of trash, compacted, petrified.

—It all makes you feel pretty insignificant, I said with mock awe.

Sylvia apparently didn't catch my tone, and said:

—Well, that's the conventional sentiment. I don't see how billions of dead animals with the IQ of earthworms diminish us. I mean, we survived, they didn't.

—You're not afraid of big numbers, I said.

—Who is?

—Lots of people. Big numbers are like eternity, we don't like to think about it.

—The numbers are too big. You can't think about them.

—Do you ever think about the world before you? Without you?

—There is no world without me, Sylvia said, and she swept her arm around, saying: —Here's the world, here I am. What other world is there?

—Imagine yourself as a ghostly soul, floating up above it.

—I do believe that when you're dead, you're dead. That's it.

—Oh, I said. That's very pragmatic.

Sylvia patted her hand on the pinkish stone on which we sat.

—This is St. Peter sandstone, she said. This is the rock on which my church is built.

Her eyes showed embarrassed delight with her pun.

—It's named for the town, she said. Not the apostle.

The sandstone was the joy of graffitists. Young lovers had proclaimed their devotion here, but one good rain could nearly erase it. "Love you Lisa" someone had scratched, and there were the usual hearts holding the arithmetic of love that yielded no sum. Also, "Anarky Now!," "Rock & Roll," and "Suck My Big Cock" (illustrated). Proud graduates had commemorated their class years. I don't know if it was for spite, or as an object lesson—I took a piece of limestone and obliterated the class of '81. Lightly I traced a heart, and inside, B+S, only in the dust.

—What are you thinking about? Sylvia said.

—Nothing. Something. I don't know.

—Confused boy.

—Yes, very confused.

There was shouting from up above, growing louder, and then three kids, two black, one white, grade-schoolers, burst out at the mouth of the path we'd followed down. They were a ways to our right, downstream. We looked. The birds who stood sentinel on the island looked. The kids were carrying fishing poles and a can of corn for bait. They saw us and stopped talking. They glanced at us a couple of times, suspiciously, as they moved away downriver, and when they felt far enough away, and safe, broke into a shouting run again. The kids were out after panfish, or maybe a whopping big carp that they would leave to rot on the shore. I thought of the trout fisherman in the river on the farm, and how thrilled these kids would be to catch even the chub that the man had squeezed to death and tossed away. Well, one's standards changed. Probably the man had once been thrilled to catch chubs.

—Why do you keep drifting off? Sylvia said. What are you thinking about?

I shook my head, and I was thinking about the night of that day on the farm, and how passionately I had declared my love to Sylvia, and had told her I wouldn't let *her* slip away. And now? And now I did still love Sylvia, but I didn't feel the conviction that I could keep her. Events and my emotions seemed to flow in their own appointed way, like the river. Like a river which is that which comes between. I could fight it but that might mean I would drown. I could drift and wind up who knows where.

—Do you think love conquers all? I said.

—*Omnia vincit amor,* Sylvia said.

—I didn't ask for a Latin lesson.

—*Omnis amans amens.*

—Okay, what does that mean?

—All lovers are demented. *Amens:* without mind. If love conquers all, it's the reign of the mindless.

—Then I'm beginning to have some hope.

—What are you hoping for?

—What am I hoping for.

—That was my question.

—Hoping for hope, hoping against hope.

—For what?

I lit a cigarette, and felt the air thickening as evening moved toward dusk, and everything becoming quiet. In the quiet I could hear the river, a low liquid chime from the shallows here, and the urgent rush of water pouring over the dam, upriver.

—You feel so far away sometimes, I said. It feels as if I could never know you. Like even if we were married and had babies and lived together for fifty years, I'd still be imagining you, and I'd be surprised to look up and see you across the table from me, or beside me in bed. It's not that you seem incorporeal, or distant, exactly.

Sylvia watched me as I spoke, then looked away, toward the river, when she said:

—Do you think you're such an open book? But it's like a judgment, you know, because secrets come from shame, and if I'm so mysterious, I'm dirty, somehow. Everyone thinks of themselves that way, because you know all your own secrets, and you think they show. I puzzle myself sometimes, I guess, but doesn't everyone, who thinks about it? Still, I don't get it. What is it you want to understand?

—I don't want to understand, understanding isn't, it's just. . . .

But I did want to understand. But I didn't know what.

—Is this a proposal, or are we breaking up, or what? Sylvia said.

She picked up a stone, half a billion years old, examined it for fossils, rubbed at it with her thumb, and threw it in the river. Maybe all this could be easily straightened out, like a simple misunderstanding in a movie which, if only someone sat down and described the situation, would just blow away, but instead leads to a baroque and mindless farce. In the end no one bothers to say just what it was that started all this. In the end it didn't matter. Someone wrote a poem and sent it to a dumb little newspaper, and the same someone happened to take cream with her coffee. If I were able to say: You see Sylvia the whole thing is this, it was just these two little clues, little scraps of nothing, really, but Carla found one and I found one, and we grabbed the loose ends like two strands sticking out of a ball of yarn, and we pulled and pulled, and untangled and untied, and we followed like bloodhounds, and when we got to the end—when we got to the end we were in the middle, and we were staring at someone holding the other end of the string, and it wasn't you, it wasn't what we expected to find, it was the two of us, me and Carla, standing in a mess of string, and you were nowhere in sight.

—Exactly what is the matter? Sylvia said. Why are you unhappy?

—I don't know what kind of world you live in, I said. I don't know if I'm in it. You seem so resigned, and self-sufficient. What do you need me for?

At this Sylvia turned to me, and her face showed genuine puzzlement.

—I don't need you, she said. I never said I did. I never would. That just means I don't need anyone. I'm always prepared to be alone. I don't want to, but I can. Oh, yeah, sometimes I want to, I imagine that would just be best.

—Why do you *want* me, then?

—Oh how could I answer that? For companionship, for sex, because of your dimples. Because you're smart, and good to me, and because I think we think alike, even if you can't see it.

But I did see it: Beauty dwells in sadness. Did I court sadness for beauty's sake?

—I do think we think alike, in some ways, I said. But I don't know if that's a good thing. I don't know if that's enough.

And then I saw that Sylvia was crying. She was sitting with her knees pulled up and her head on her arms which wrapped her knees, and outright crying as if something had suddenly given way, and I saw her tears fall and splotch dark on her light cotton dress. I reached toward her but she shook her head, not lifting it from her arms or looking at me. Then I didn't know what to do with my hands, and I rubbed at the sandstone that came away in grit beneath my palms. I looked down and saw that I was rubbing the big cock, and I pulled my hand away, horrified, and I rubbed my hands furiously together as a shudder convulsed my chest.

—What, what is it? I said, but there was an edge of irritation in the compassionate concern I had meant to convey.

She just shook her head again. I tried hard to concentrate, to force a form on what I felt. Then I thought I had it. I spoke, haltingly:

—You've been holding on to something for a long time. I think. Clutching tight to something, just to keep your head above water. I don't know what it is, exactly. Your knack for survival, or what you've survived. Why does it have to be this way? You're so smart, you're so wonderful. It's some kind of idea. It's what keeps you going, you think. You think you need it. No: you do need it, everyone does. You did need it. But if you're holding so tight to that, can you hold on to me, too? Can you fit someone else inside this idea?

I touched her shoulder. She didn't raise her head. Her crying had quieted.

—Maybe I'm wrong, I said.

I drew my hand away.

—I did something really wrong, I said.

And she suddenly stood up, and said,

—Don't say it. Don't say anything more. I'm leaving. Don't follow me.

She jumped down from the sandstone ledge and passed in front of me as I reached out, saying,

—Sylvia!

Sylvia!

She looked straight ahead, her face streaked wet and angry. She clattered through the limestone scree toward the mouth of the path. I went after her. The path sloped up sharply here and she was almost on hands and knees, kicking up dirt and holding on to exposed roots as I came up behind her.

—Sylvia!

Sylvia?

—Sylvia, wait! Why are you—

I reached out and grabbed her ankle. She kicked it free. She turned in a crouch and swung back her hand and she hit me, hard, on the side of the head with an open hand. I jerked back as the pain shut my eyes. Then I lunged forward and swung at her, but my fingers only grazed her ankle. As I fell forward her foot shot back and her heel caught me solidly on the lower lip. I felt the instantaneous swelling, tasted dirt and rubber and blood. I landed flat in the dirt and rotting leaves, then slid slowly down until my feet touched the pile of loose rock. I looked up and saw flashes of her dress through the trees as she ran up the path.

I went back to the ledge where the remains of our picnic were scattered. I sat down and took a gulp of wine, and it stung my split lip. My jaw hurt too, where Sylvia had hit me—the brunt had landed just in front of my ear, where the jaw joins the skull. The kick I think was not intentional, she had lost her footing. I hurt in many places, and I felt a sick hollowness in my chest. But there was also some sense of relief, as if a dreaded necessity had been dispatched. I am always prepared to be alone, Sylvia said. And I was, too. But we didn't know very well how to be together.

Now the bank behind me blocked the sun, and the river gorge filled with darkness from the bottom up. The heron and the egret still held to their separate ends of the island, and I watched them to distract myself from other thoughts. They were turned toward each other now, and in the dusk I could imagine that they were looking at each other, and perhaps they were. Did some wading bird telepathy pass between them, a current carrying ancient intimations? Maybe they were in love, smitten with a wide-winged, awkward love, and doomed to disappointment by the accident of species. And at that moment, as if it knew my thoughts and couldn't bear the truth it read there, the egret hoisted its big white wings and pulled itself ponderously into the air and took itself across the river into the lower branches of a dead elm tree. The heron also rose up with a great commotion and sailed away downriver. And just as the air grew quiet, from the island came a raucous cawing and an explosion in the spindly poplars, and one after another a half-dozen barrel-bodied, broad-winged birds rose up, winding in a ragged spire up through the gorge, the phalanx lengthening and widening and rising like a thread pulled up in a whirlwind, until above the highest trees the gyre broke and they flew out of sight.

—They make an awful racket, don't they?

I turned sharply and saw an older woman in blue slacks and a gray sweatshirt, with a walking stick in her hand.

—Black-crowned night herons, the woman said.

I nodded and found voice to say,

—Yes, I've seen them before.

The woman wished me a good evening, then went away up the path. I had been hunting for the name of those birds. Night herons. I had seen them often on the lake where my mother lived. It was eerie how every evening they burst up of a sudden from where they were hidden in the trees, their loud voices like coarse bells clanging. You could never quite predict the moment, it was always an unnerving surprise. It wouldn't be long now before they flew south to winter. I thought of the great currents of life that swept the globe, the unseen highways of air, the mindless paths and patterns of creation. There is so much we can't see.

The egret perched ghostly in the tree across the river. Why not say it was my father's ghost, which lingered here so near where he died. I thought I ought to leave, and I was scared to be down here in the dark. Obsessively my mind rehearsed the confession to Sylvia that I hadn't made. It was essential for me to understand that the betrayal was real, there was nothing hypothetical about it; it was something I was capable of because it was something I had done. And something I wouldn't have thought I could do. And so I was not who I thought I was. But of course I could not believe that.

I did something really wrong.

I made my girlfriend cry and she ran away.

My girlfriend was always running away.

I sat one billion years deep in the earth. The cliffs displayed the fall of all that rose. Trash heap, junkyard. We mindless living whistling atop an awful, magnificent graveyard. One billion years. The big numbers scare us; we cannot hold them in what we have of mind. I sat on the crest of time, as everyone does. Why spend one billion years and so much stuff to purchase this, an emptiness at the bottom of the river gorge? No reason.

Where is the beauty that dwells in sadness? There was no solace in this death-walled pit, or in the eroding stone, the bedrock of Sylvia's church of oblivion. Maybe we didn't really think the same, after all. Then I felt almost priggish as I stuffed the remains of our picnic into a bag to carry it out. I took one last swig from the wine bottle, and poured it out to darken the sandstone.

I started walking down the shore, to try to clear my mind, but there wasn't a good path, and it was becoming too dark to see. I stumbled on rocks and branches, and walked right into a creek's mouth or a seepage puddle, and panicked and thrashed across, soaking my jeans to the knee. I heard voices in the trees, and I was frightened. Down the shore I saw a fire, and another across the river. I was in the wrong place. I was somewhere I didn't belong, and I had to get out. I had come too far from the original path, and I thought I heard voices back there, too. I found something that looked like a trail going up, and started running up, until a branch raked my face, and I stood blinking away tears and looking for the trail that wasn't there. I thought *jesus christ what have I gotten myself into.* I was doing everything wrong. How could everything have gone so wrong?

I continued up, though there was no path. My arms and neck became badly scraped, several times I fell hard to my knees. Somewhere I dropped the bag of trash. At last I reached the top, and lunged through the last line of brush, and almost knocked over a cyclist riding by on the dark path. She shouted and cursed and rang her bell, and I gasped, and I was nearly mad with fright and frustration. I wanted to scream and run in circles till I dropped. Instead I walked back toward the car, breathing deeply the cool damp air. Crickets keened in a minor key, and I couldn't believe that anything could be good again.

The car was closer than I thought. I stood across the road from it, and gave it a long, long look. I remembered Carla's saying that the car was my destiny. And I considered leaving it here and walking away as Sylvia had done, we would both walk away and leave whatever destiny resided in the old blue car to sit and rust and be towed away. And I stood and looked, and then I said:

—No.

Said it aloud for whoever might hear. Said it loud for the trilobites and the river-bottom denizens and the crickets and the damned blue car. And then I said:

—*Fine.*

And I threw down my hands and shook my head, because it had all gone too far. And I thought: All right, then, the world is a graveyard, and love conquers nothing and all lovers are demented, but that is the situation, and if that is destiny then I'm not going to park it I'm going to *drive* it. Enough of this damned diffidence and brooding and imagining. I would *do* something, by god, I would start with myself and see what might follow. But what to do? For starters, I could leave this place.

There was solace in movement, even though I didn't know where I was going. I took clues from the car. Remember that first drive, when the Bel Air resisted turning off at Sylvia's exit, wanted to go to Iowa? Tonight I offered it the choice of driving to Sylvia's, and it firmly declined. And I hatched a plan: I would never cross the river again. Say that's the past and I will dwell on it no more. Nothing specific in that, no conclusions about Sylvia, but a scheme to limit my sphere of action, give bounds that would make decision easier. There was a whole half a continent out here, the whole great West. That should be enough. Why look back? Lots of people lived their whole lives and never crossed that river, one way or the other. Of course I wouldn't follow through on it, but what if I did? I would become a true eccentric, I would be defined entirely by this one choice, I would be The Man Who Wouldn't Cross the River.

The river road brought me into downtown. I drove past Frank's, and his lights were out. I drove past Carla's, and her place was dark, too—she would be at work, cooking something wonderful, cooking wholeness to a turn. I wound around through downtown a bit, but the stoplights were annoying. I found myself on a broad one-way street going south out of the city. The lights were timed, so if I kept a steady pace they turned green—green—green ahead of me, and I cruised at thirty-five miles an hour as far as the lights led me, thirty blocks south, through the blighted, and then the threatened, and then the marginal neighborhoods, to the momentarily stable neighborhoods at the city's southern bounds. There the street turned to two-way traffic, and I turned left two blocks and got on the south one-way's northbound twin, and followed the lights again back into the city, traveling back to deterioration and compression, and then back onto the southbound, and I drove this several times around. There were east and west one-ways that went through my neighborhood, from the lakes in the west to the river in the east, and when I tired of the north-south, I drove these several times, stopping perhaps once on each leg to draw in sync with the lights, and toward the end I managed it—acting, I would say, almost wholly intuitively—to swing around for the return trip with such refinement that I did not have to stop at all. The car liked it. It had never run so smoothly, responded so smartly. An outside observer might have said I was going around in circles, but I knew otherwise.

I drove this way until it was good to stop. I didn't listen to the radio. There was traffic but I attuned myself to it. I cut no one off, never used the horn, did not allow myself to be hemmed in or be forced to change lanes suddenly or accelerate or brake with any urgency. I drove with the window down and felt the cool moist wind soothing on my prickly scratched face and arms. And about the only thing I thought of was my driving instructor back in high school, and how proud he would be of me. My driving instructor was the twin brother of Boscovitch the ranger, who was not dead and did not sleep nor ceased from vigilance. —*Aim high in steering,* I remembered him saying, and though I couldn't remember anymore what that meant, it seemed almost unbearably wise.

The journey ended on the westward leg, and I flowed off the broad way into narrow residential streets near the lake, and onto the lakeside boulevard, and came to rest before my mother's house. There were lights on, but they were the lights my mother put on when she went out—who but me would know the meaning of those lights?

I went up the stone walk toward the front door. It was riverstone, it was packed with fossils, I felt them under my feet. I remembered my childhood fascination with those magical, ancient shapes in the mundane context of the front walk. Brachiopods, I thought. And I guessed that this would have been one of my first intimations that the world was full of wonders. I had forgotten it.

I went in the back door. The house welcomed me. Each of its myriad parts whispered greeting. Tonight the world was everywhere animate. Each floorboard, every splinter and grain of each floorboard, and every nail, and every thread of carpeting, and every turn of every rung of the balustrade, and the toaster and refrigerator and every inch of wallpaper and every piece of furniture and all their component parts, and the bar of soap by the sink, and the soap that had washed off the bar of soap by the sink and dried to become the soap that coated the place where the bar of soap by the sink sat—every molecule and atom down to the indivisible bit at its heart of being, if there was such a thing—all, individually, and as one, welcomed me; and I was overwhelmed. What it all said, with quiet certainty, over and over again, was: *home.*

. . .

I worried about madness. My father, I had to conclude, had been insane. This communing with the home-soul was not a projection of my memory or desire; the house *was* alive. So I believed. This feeling was not confined to the house, I had known it at the river, too: the river and its banks, the island, the birds, and all the dizzying specifics of that place, of all places, lived and spoke. I heard it a certain way because it was I who heard it, but it was not I that spoke. If this was a warning sign that the flaw of madness, hidden in my genes, had begun to manifest itself, it did not worry me. I didn't feel paranoid or oppressed; I was grateful to be able to hear it. That I thought of the car as a sort of spiritual guide, sometimes friendly, sometimes sinister, worried me a little, but I did not dwell on that now.

I went slowly through every room in the house. It was like meeting long-absent relatives whom you like. It wasn't gushy or affected—only the living room, which still held some of my grandparents' overly ornate furniture, threatened to make a scene, as of too ardent hugs and slobbering kisses and exclamations on how big I'd grown and why didn't I cut my hair.

Downstairs was aunts and uncles and older cousins. Upstairs: the landing / hallway, the favorite aunt, the one you could confide in, your mother's youngest sister; the family room, closest cousins, who are like siblings but better because they aren't; bathroom, your beloved grandfather, who got so old, and became not himself, it scared you and you didn't understand, and you went on fiercely loving someone who wasn't there. And the bedrooms—the bedrooms thrillingly, dangerously specific. My brother's bedroom: my brother indeed. My troubled brother with whom I rarely spoke now, not because we didn't want to, but because it had become so difficult. My brother whose one desire was not to be my father, which very desire and all effort to attain it ensured its contrary. My room: oh, hello, Bryce, hi there, young Bryce! It wasn't me. It was that earlier me whom I was and was not, whom I was able to greet warmly, and consider with some charity. The overly serious, uncertain, slightly arrogant young Bryce. What remained of him, and what had changed, did not concern me at that moment—only that we could meet here, easily.

There was the attic, which I did not enter, though it had convenient stairs. The attic was relatives whom you'd maybe never even met, and you knew exactly one particular fact about them: the great-uncle from Great Neck who was gassed in the Great War; the aunt who ran off with her dentist, an Italian; the second cousin who won the lottery, and the one who miscarried four times. These carried mystery and persistence because the evidence of their existence was so striking and so spare.

And then the heart of the house.

I hesitated at the door, while young Bryce shoved past me and bounded into the room and flopped in a sprawl at the foot of the bed where his parents were still sleeping on a Saturday morning, and bounced three times to arrange himself, and announced in a loud, cheerful, challenging voice what he was doing today, where he was going, and with whom, and he'd see them later, maybe by dinner time and maybe not, and receiving only grunts from his father and nothing at all from his mother who was awake but pretending to be asleep, then sprang up from the bed as if ejected by a spring and barged past me and loudly down the stairs, leaving the dark door open.

Have fun, Bryce.

Cautiously I entered. It wasn't Saturday morning, no stream of summer sun. It was dark, the bed was neatly made and empty, but the residue of souls lingered on it, and on everything in the room. I did not turn on a light. The essence of this room was its dark warm mystery. I went to the bed, with its blue chenille spread, and sat, then lay back. The mattress gave and the springs muttered. It was far too soft, I could never have slept comfortably on it. It was perfect, it was my parents' bed. Above the dark, simple headboard hung a picture, a large framed photograph of me and my brother as children. Guessing, I would say I was two or three, and Colin five or six. It was a formal portrait, against a pearly backdrop. We were seated on a table, probably, and Colin sat behind me and sort of propped me up. I had on striped overalls with a fuzzy bear appliqué at the breast, and a white sweater; Colin was natty in a little plaid blazer, white shirt and tie. My head looked large and rather blocky, my cheeks still baby-chubby; my eyes were open wide, wide, and my mouth was open in a happy expression that couldn't quite be called a smile. Colin already had his shy, uncertain smile. He was thinking about something else—he was old enough to want to be someplace other than where he was, and I was young enough to simply respond to where I was. I was a sucker for the attention lavished on children to get them to look nice for the picture. Maybe Colin was even thinking what a sucker and a ham his little brother was. His face had already taken on the angular shape it would show later, whenever his worry made him thin; it showed a clear resemblance to our father.

But what sweet faces. Such bright eyes, and golden hair (though the photograph was black and white). Those souls lingered, too. These few years were nothing to them.

This was the room of things as they are and ever would be. I felt lucky to have such a place, and knew that not everyone did.

I stood and went to the dresser. The dresser was finely made of dark walnut. The drawers slid smoothly open. The drawer pulls were brass rings worked so they looked like wreaths, and they jingled a little after the drawer was opened, dispersing the drawer's momentum, and they jingled in a different, a closing way, when it was shut. I pulled open the two shallow top drawers. These had always been called the junk drawers. The dismissive name was camouflage: the code of our family life was scrambled in what had been thrown in here with seeming carelessness. My mother had one, the right, my father's on the left, but they mingled and overlapped. What was in there really was junk: pens that didn't work, printed with the names of long defunct companies; keys without locks and locks without keys, all, mysteriously, very small; calendars from insurance agents which are supposed to fit in your wallet but are a little too big; business cards where the printed address has been scratched out and replaced with a new, handwritten one, illegible; a monogrammed handkerchief, the initials didn't belong to any of us; a red maple leaf pressed between waxed paper; a piece of dirty, broken pottery, it was unearthed somewhere but was not very old; marbles, all aggies and rather homely; empty keychains, the kind that pop open unexpectedly, the kind that dangle a company's or sports team's or school's logo, encased in cloudy plastic, and that part is gone; clip-on earrings, one of each pair; a package of single-edged razor blades; a bar of soap from a New York hotel; scraps of paper with addresses and no names; two fuzzy snapshots, one of a suburban house in a winter of big snows, heavy winter light of late afternoon pressing down, and one of a small sailboat, a bright afternoon, but overexposed so the image is dark, and the frame is tilted so the boat races toward the upper right corner; a shopping list; dog tags of a soldier, and dog tags of a dog. And more.

I looked in the drawers but didn't touch anything. Young Bryce, when he was alone here, would rummage through the drawers, and sometimes steal things, an Indianhead nickel or a Mercury dime—he was out for lucre, not sentiment (but didn't he feel the power there?; he did, and he stole to deny it).

Of any one object in the drawers, I could not say what it meant. All together it meant that things were built up as well as fell down, and against expectation, against probability, some things stayed built for perhaps long enough. The world-soul was in the building up, as well as the falling down. With all that had happened, with how my father had left us, the cycles of life revealed in this miscellany might have pointed to tragedy, and sometimes did. But the specificity and intimacy of these things spoke also of a solid sort of peace, a blessed belonging.

My father's junk drawer was the only space of *his* that had not been cleaned out since his death. How could you clean out a junk drawer? And so, like the arbitrary sounds of a language, these broken, useless or never-to-be-used objects could speak of him with utter clarity, and say what he was most simply, most truthfully, with utmost dignity: one man who had lived and died.

I thought of those tiny stone cogwheels Sylvia had shown me—crinoid rings, she called them. Such delicacy at the bottom of the death-walled pit. Why save such intricacy for a billion years?

There I was in the mirror above the dresser, there in the intersecting light of the stairway chandelier and the lakeside streetlights. My face dirty and scratched, and hair every which way. And there I was in telescoping images from dirty-faced, tired Bryce at twenty-five, down through countless, shifting avatars to the bright-faced, fair-haired, gaping boy in the photograph. This resonating current of lives was not entirely illusion—I could see the photograph of the children, in the left-hand side of the mirror, slanting away.

I went into the bathroom and started a hot bath running.

In the homey old bathroom, white tiles trimmed black, comforting clunky old fixtures, I pulled off my grimy clinging clothes and wondered why lately I was always so disheveled. I seemed always to be forcing against resistance, pushing pull doors.

I stepped into the tub with the water still running, splashing, steaming, and lowered myself in, and slid slowly down till I was completely immersed, head and all, holding my nose shut, and came up streaming and breathing deep sighs. The scratches on my arms and neck stung from the hot water, but the stinging subsided. My knees were scraped and bruised, I now noticed, as were the heels of my hands which shed a thin wash of blood into the water.

When I was with Sylvia I always hurt myself. An offhand remark I might have told her as a joke. But what truth might there be in what simply seemed to happen, or more important, what meaning, in these patterns that caught our notice but evaded deeper scrutiny?

Well, what? And did they evade deeper scrutiny, or did I just not trouble to delve?

Well?

The cogent question can make one seem wise, without risking the deflating commonness of an answer. I felt safe here, and happy, if only from the physical ease of the soothing hot water. I could risk an answer, if only to myself, so I turned my mind to this thing which I had been thinking of as *the secret desire.*

One's being has a goal; one doesn't know what it is. But by paying attention to how one reacts to various events, even trivia, minutiae, one may begin to discover it, begin to plot points and draw lines and in time have a map. I thought of all this resistance, the psychic scrapping that centered around Sylvia. And of the drastic mood swings of this evening, from the elated love at Sylvia's room, to my churlish unease by the river, to the calm of the drive, and the warmth, sadness, and seeming clarity of this visit home. But with Sylvia—did I hurt myself because of her, or because I resisted her, and drew away? I needed more coordinates, more landmarks. And then, I knew that Sylvia would be but one region on the map of my secret desire. The secret desire is not identical in all people, like instinct in animals, but it exerts as powerful a force; it is the uniquely human instinct.

All one's tastes, ones likes and dislikes, how one reacts to everything, these are determined by the secret desire. The secret desire is compact of genetics, environment, how one sees oneself, what one fears, what one loves.

This was soothing, and seemed to make sense as I drifted in the rising steam. You would never fully understand your secret desire. It would not be revelation or salvation; the secret desire itself is not *the answer.* But you might recognize its direction. You might not *like* your secret desire. But you couldn't change it. If one recognized one's secret desire, might it stop one from jumping off a bridge? It might push one off. I had to consider that my father had fulfilled his secret desire precisely by his death. Of course, circumstance acts to channel the secret desire; if a river is dammed it will try to find another way.

Thinking only of this evening's events, it was easy to list some components of my secret desire: I wanted closeness, and sex; I wanted to avoid conflict, but not at the cost of surrender; I wanted anonymity, oblivion, to be nowhere; and I wanted a home. Safe in this steamy fastness, I believed that this last was the heart of it. Who didn't want a home? I had always been drawn to other people's families, no matter how strife-racked, contentious, or, in Alison's words, plainly fucked up, as long as they seemed like real families, bound by whatever fraying thread. Why hadn't I brought Sylvia here? I hadn't even thought of it. Why hadn't I brought her to my home and my family? Only my mother lived here. That didn't matter, numbers didn't count here. Sylvia had shown me her family. She hadn't exactly brought me to it. She had presented it, maybe, as evidence. Like my own justification by blight. Maybe she couldn't draw together the flown and shattered bits of her life. And maybe I could. Maybe I would. When you are growing you turn away, and when you are grown you turn back. The passage is a circle. That is why circles are not futile. You arrive at the same point, but the point is not a point alone in space, it is a point in the circle, and there is a difference. The world is a round world, a sphere indeed. The secret desire, then, is perhaps not a map but a globe.

One is given chances, and the outcome is determined by one's actions in relation to the secret desire. One gains by according to the secret desire, and loses by opposing it. I wondered what chances I might have recently lost, and what remained. I only feared that when I saw Sylvia again, if that was a chance that remained, all this calm certainty would fly away in the usual storm of emotion. If I decided I wanted to love and protect Sylvia, but could not act on that wish, was that my secret desire speaking? I think I wanted to bring Sylvia to a safe place; but then why was I so awful to her?

I lay warm and wet and sleepy, and was growing greatly wrinkled. My mind registered a familiar, comforting series of sounds: a car in the alley, pausing, idling; the garage door going up; the car driving in, the door going down; the sticky door from garage to back walk springing open with a shove, a nudging kick with instep and knee; high heels on the walk; the back screen door, keys in the lock; the whooshing compression of air as the inner door opened; the suddenly amplified sound of the same woman's step in the back hall, coming into the foyer, dropping her purse on the foyer table, kicking off her shoes, sighing. Every light in the house, except for in her bedroom, was on. Did she notice? A fleeting scent reached me: was it possible, from all the way downstairs and through the thick air of the bathroom? The inimitable scent of been-out-and-just-come-home mother. Perfume, night air, dry-cleaned clothes, hairspray, powder. If I tried to smell it it fled. I steeped in a thoughtless, steamy cloud, transported far away from my surroundings, but something elbowed the ribs of my contentment. I sat up in the bath—

—Mom! It's me! Hi, I called.

She would not be expecting to find me in her bathtub. I hadn't been over to the house in weeks. The soft sound of her stockinged feet coming up the stairs—I could imagine her craning her neck upward as she mounted the steps.

—Bryce? Where are you?

And she came to the bathroom door, saying,

—What on earth?

I leaned over the edge of the tub to see her.

—Just came over for a bath, I said. I tried the kitchen sink, and I tried the laundry tubs, but this is just right.

She entered and sat on the toilet seat cover, and leaned forward to kiss my forehead.

—Well, Goldilocks, nice to see you.

But from how her smile fled it was clear I wasn't so nice to see.

—Good lord what happened to you? Were you in a fight?

She scanned for more evidence, saw my pile of muddied clothes on the floor.

—Bryce! What's going on?

—Nothing, really, I'm all right. Let me get out of here and I'll tell you.

I studied her face. I was still glad to see her. Her dark straight hair was showing gray, which she sometimes colored and sometimes didn't. It reached almost to her shoulders, and was cut to one length, so that it was swept back from her fine brow, revealing a straight, firm hairline that to me spoke intelligence and poise—so I immediately liked anyone with a hairline like that, if the forehead was up to it. She wore a gray silk dress—she had been at some kind of fund-raising dinner, she said. Her make-up was showing a long evening's wear, and the way it crinkled around her eyes and mouth was enormously touching. She was fifty-two. I regarded her perhaps a bit too glowingly. She said,

—What's with you? Do you know what time it is?

I didn't. Almost one, she said.

—Geez.

—Geez. Don't you have a bathtub?

I said I'd just been out for a drive and had decided to stop by. She didn't know I had a car. I said things hadn't been going well with the woman I had been seeing. She'd never heard Sylvia's name. The car, the woman—I had to get out of the tub before I could explain it all. The water was now grimy and tepid; I had never even gotten around to washing.

—I think there's still some clothes of yours here, she said as she gathered up my clothes strewn about the floor.

When she left I quickly washed and rinsed under the shower as the gray water drained. I wondered at how one could be so happy, so understanding, to find an uninvited guest in one's bathtub at the end of a long tiring night. A line of demarcation on the map of her secret desire. I considered that mothers were less aware than anyone of their secret desires, yet acted most truly to achieve them. Not to say that this always brought universal acclaim.

She had found a high school T-shirt and a pair of jeans—my most treasured pair of jeans, in fact, faded almost white and patched with small squares of various fabrics from below the knee all the way up the thigh, and on most of the seat. That had been the epitome of cool in high school, and I had worn them through one year of college before they became an embarrassing souvenir of passe tastes. Now they felt like something rare and wonderful made especially for me, without my asking. Which they were, in a sense, made for me by another me. I could bring these fragmented selves together. I was the one who could draw together things that were apart. Maybe.

I had a drink with my mother before I left. I told her a bit about Sylvia. She said she'd like to meet her; I said I didn't know if she would have that chance. I agreed to come over and help with some yardwork in the next couple of weeks. Then I went out the front door, and down the fossil walk with my dirty clothes in a sack under my arm.

Disappeared

I will not cross the river. A flippant, fleeting notion at first. But then, why not? You can make simple decisions like this, for no apparent reason, only to give an arbitrary limit, impose some kind of discipline. So I did. I cautioned myself to remain in control of the discipline, and not let it rule me. Madness is allowing arbitrary rules and omens to dictate your life entirely; sanity is allowing arbitrary rules and omens to dictate your life just mostly.

I drove, and contemplated the secret desire. For days I did little but drive.

I drove within the one-way loops I had discovered. I drove

along the west river road, to test myself: one simple movement of the wrists, ten seconds crossing, and it would all be wiped out. The tension between the time devoted to this exercise and the simplicity of ending it became excruciating, and thrilling, the longer I maintained it. It was like plying greater and greater force to the end of a lever wedged beneath a great stone. But the stone, as I pictured it, was small; a tiny, immensely weighty stone, like certain objects found in the depths of space. I imagined that this confinement, this tension, was giving me power. It was a weird sort of meditation. It struck me as the essentially American meditation. We don't have mountaintops, bamboo groves, desert anchorages—we have roads and highways. For some reason I wanted to be essentially American. I had never thought about it before. The car had something to do with it; it was an essentially American car.

I didn't feel like talking to anyone, or being around people at all. It wasn't an aversion, just a sort of human apathy. My copy for the paper was scant, because I avoided interviewing sources for stories. I avoided the office generally, and left copy on Lillian's desk late at night.

Often I wore the old clothes I had picked up at my mother's house. The jeans more patches than original material; the T-shirt gray athletic department issue—*Property of, XL.* These clothes seemed to embody some kind of simplicity. They were relics from a sunny, golden time, an innocent time. Which was a lie. It was in my last year of high school that my father killed himself. I was aware of the lie, and I indulged in this small artifice willingly. I described the lay of desire thus: some people imagine the sunny golden time in their past, and live nostalgically; some imagine that it is yet to come, and live with yearning for a messianic age. Both were delusions, but I preferred the former, for at least with nostalgia you weren't always waiting.

I wanted to know everyone's desire, and I couldn't stand to be around people that I couldn't read this way. I couldn't tell about Carla, so I didn't go to see her. Frank I understood. His dwelling place was a third category, which is the golden moment. So I could be with him, and sometimes I went and sat in his studio and watched him paint. We hardly spoke beyond hello and goodbye and how about a beer. Frank saw that something was going on. He waited to see what it would be. He wasn't one of those people who always want you to talk about it.

With this solitude, restriction, and analysis there came a sense of detachment which I knew was dangerous. It drew me away from the ordinary life that I was trying to draw Sylvia into. To counteract it, I replayed the scene at the river with different endings. I take her in my arms, I comfort her as she cries. We speak all our secrets; for us two we forge one desire, one home. Before, I had thought that the things that were understood did not need to be said, but now I saw that was wrong. Because the things that were understood were perhaps understood differently, they had different meanings for the two of us. And also, the things understood were so hard to put into words, even in my head, it made me think that it was necessary to describe them, or at least indicate them, to each other. Point: see that there. What about it?

I almost quit drinking entirely, and because of this I realized I had been drinking a lot. I smoked a lot of cigarettes as I drove. I drank a lot of pop. I stopped at gas stations and convenience stores for supplies—cigarettes, pop, gum, snacks—and these breaks felt like stops on a long journey. I drank the pop of my youth: Mountain Dew, Orange Crush, Dr Pepper.

. . .

I couldn't sleep at night, so I drove. What a world I found when I rose from my restive bed and went out to drive through the night. The world where American loners and misfits drove and drove, and stopped the car to kill. At convenience stores as I loaded up on pop and junk food, I wondered if this appalling diet of sugar and MSG would turn me into a homicidal fiend, as had been known to happen. Many of the sparse cars on the road in the middle of the night were old junkers like mine. Rusting leviathans that belched blue smoke, the American dream gone to nightmare, that inversion so often noted. With vehicles that big, there was something monstrous in their decline, when their grandeur had come from bigness alone, and the gracelessness of the design glares through as the shine wears off. The darkness was these beasts' domain—particularly the hours between one and five a.m. While the working world slept the sleep of the respectable these vehicles of fear and subversion convulsed the night with the raging of perforated mufflers, corroded tailpipes that raked the pavement, trailing fire through the streets. Through *your* street. If you woke one night at 3:17 to hear that nearing roar, you held your breath against your heart's terrified crescendo until it passed. As it faded past someone else's house, into someone else's neighborhood, you laid your head down, touched your love's warm shoulder, returned to sleep's dream castle of security. What if it didn't pass? New Yorkers, Impalas, Fairlanes, 500s and Furys. How do you like your planned obsolescence now, Mr Ford? All it took was a few spots of rust, a strip of chrome askew or missing. This Caddie won't go to the country club no more. Now it was a dangerous power, a weapon in the wrong hands. But then, wasn't it always? These cars were symbols of America itself, our benighted sense of history, how we failed to see beyond the shining metallic instant of creation, the latest big idea. I myself wielded one of these monsters. Before, it had only ever gone to church. Still, compared to an LTD gone bad, the Bel Air was relatively benign.

One night at an all-night convenience store, as I stood at the cooler deciding between Mr Pibb and Tahitian Treat, a man and woman came into the store. A repulsive couple, unkempt and fat and loud, and very drunk, both of them. They started to argue—whether to spend these last three bucks on beer or cigarettes or gas. But the substance of the dispute was quickly forgotten, because he was a shit prick and she was a fucking cunt, so they respectively declared. They could hardly stand up, and when she shoved him, he didn't. He staggered and slammed into an instant teller machine and slid hard to the floor. He was up with surprising quickness, and as he stood she was at him again, but he threw her back into a display of little bags of chips. They were right in front of the counter, near the door. They were nearly the same size, and with them both so drunk it was a fairly even match. The clerk (a middle-aged Laotian man working two jobs to support five kids—we were by now familiar acquaintances) was yelling at them, and they were yelling louder. I was on the far side of the store. I was scared, I was trembling, and wired. I watched them throw each other back and forth. I was sickened and exhilarated. The clerk was hysterical, his English went all to hell, and he screamed at the combatants in Laotian; his meaning was nonetheless clear. I made my way along the walls, waited for the battle to flow toward the back of the store, then slipped out the door. I jumped in the Bel Air and drove away fast.

Then I couldn't stop thinking about that incident, though I didn't want to think about it, I didn't like what my reaction seemed to say about my secret desire—the sick thrill it had given me, as if it had been some vile entertainment. It had. A squalid novelty. It forced into my mind the scene at the river with Sylvia, where emotion had spilled over into real violence.

She had hit me and she meant to hurt me, and she had. I still had sharp pains when I ate, from the blow to the jaw; I still tasted raw flesh on the inside of my lip, where it had been smashed against my teeth. I had struck at her too, and it was only incidental that I hadn't hurt her. What if her face had been within range of my maddened hand? What if we were poor, and without hope, not born into privilege? We were not two drunks in a convenience store. We would never be down to our last three dollars. We were not those people. We never would be. No, we would not. Also, we were not *we.*

My infidelity dwindled to insignificance. I still felt shame and guilt over it, I panicked periodically in fears of disease, paternity. But mostly I had thought of it as a cause for Sylvia to end us. As if there existed some kind of contract between us, to be voided only with just cause. There was none. I had no claim on her. She could tell me to go away for no reason at all. If there was no contract there were no breaches. Why was I thinking so legalistically, when the *point* was to live intuitively, to follow the secret desire? Because with reason you can reason yourself out of it. But even if she doesn't know, and even if it doesn't matter anymore, for Sylvia and you, you know what you did, don't you, Bryce? I know. It wasn't a matter for reason; it was a fact to embrace in the twilight halls of unreason.

The other side of the river, which I denied, to which I would not cross, became the focus of all my thoughts and actions. You can make an artificial, enclosed system like this, and use its structure to draw in and contain the real concerns which have prompted it. You can build the same solid fortress and use it to exclude and insulate. The crux here was fine, like the wrist-twist difference between this side of the river and that. Between a lover and a victim, tenderness and murder.

When I drove at night I thought of Sylvia walking her river bridges. I followed her in my mind, starting in her room, and took her down the fire escape, toward the river, everywhere she went, to keep her safe. The rules of this imagining demanded relentless specificity, and killing concentration. I had to stay with her every step of the way, no quick cuts or time-lapse allowed, and I had to follow her from the moment she stepped out of her door to the last step of the fire escape on her return. Always something drew my attention away before I had completed the walk with her. But I could always start over. It was my game.

I thought of what Sylvia had said when I tried to persuade her to stop her night walks: *You can't change me.* And how the secret desire had whispered: *Just watch me.* And if that was its voice, then it was a violent machine indeed.

Another night, another convenience-store stop. Only as I came to the counter and saw the clerk did I realize it was the same store where the fight had taken place (the nighttime city was another wilderness, without demarcation; all the streets, all the neighborhoods, all the convenience stores were one). He remembered that I had been here that other night. He smiled broadly, shaking his head as he recalled it.

—Crazy, he said, twirling a finger at his temple.

I laughed with him, agreeing—crazy. What got into people? But I felt somehow ashamed. He stood here night after night, and such things, or worse, could happen on any shift. And I, I had run out on him, to go drive in circles. What could I have done? That didn't matter. I should have stayed, I might have been of some use, if only to calm my friend down once the battle was over, give him someone to talk to once he regained his English, or listen to him ramble in any language, what did it matter? He didn't hold it against me. He was even happy to see me. I was just a familiar, harmless face.

Then I left thinking about that couple, that night, thinking that I stood too much on the sidelines. Well, I was a reporter, that was my job, my training, my temperament. What shit. Harmless. Was I harmless? The night when I witnessed the fight, I had tried to see my commonality with those people, recalling the real, vicious rage with which Sylvia and I had struck at each other. But it hadn't sunk in. It was a pallid, there-but-for-the-grace-of-god triteness. I had tried to see the sameness, but I had focused on the difference, that thin line of grace, if that was what it was. What I saw now was: You're just the same. You're not drunk, impoverished, brutalized, but you're just the same. You don't act as they do, but you're just the same. I told myself this. And I resisted believing it. Who wouldn't? But if *those people* are your fellow creatures, if you share a world, how can you deny it? To sit and judge from an abdicated jurisdiction, who hears your verdicts?

Harmless? Hardly. Sylvia had said that her mother thought herself harmless. That first day at Gran's when we had sat together hidden in the long grass, Sylvia had so clearly revealed what harm the innocent could do, yet I had thought us protected. Protected from the evil world of grown-ups. That was the world of which I wanted no part. But look: there is no other. And love won't make it otherwise, nor blindness, nor denial. Then the answer was simple, obvious, unavoidable: grow up.

. . .

I crossed the river. It was easy. As soon as I did, I felt stupid and childish to have wasted so much time. I went over in the evening without calling. Although I now glimpsed what harm I could do, although I now saw the need to forgive and ask forgiveness, still I went to her with heart unpure and troubled by pride. Why was I always the one seeking Sylvia? Would she ever come looking for me?

I parked on the sidestreet and walked around the back as usual. The space where Sylvia usually parked was empty. I bent over the muddy ruts and peered like an old tracker, trying to see if the car had been there recently. But I don't know what exactly I was looking for, and Sylvia wasn't the only one who parked there. I saw tread marks.

I went up. It had been a long time since I had bothered about stealth, and I had come to understand the scofflaw attitude of inveterate criminals: when you have broken a rule or rules for long enough without consequence, it ceases to feel as if you are breaking a rule; you would be surprised and maybe outraged if you were caught and punished.

I peered through the window. The bed was stripped without even a mattress pad, just the blue-striped blue-buttoned ticking. I looked through the right window, I looked through the left: the same emptiness from both views. I knocked on the window, quietly, so as not to startle the no one there. I looked through the windows a long time. So long that the arrangement of the objects in the room began to take on a weighty, destined feeling. The chair how it was pushed away from the desk looked not as if someone had just got up from it, but as if it were arranged to look as if someone had just got up from it. The intimate objects on the dresser—skin cream, hairbrush, tiny woven basket with one barrette protruding, silk jewelry box—looked like someone's idea of the personal possessions of someone who lived in a room like this. Like a painstaking historical re-creation. This is the room where. . . . The room as it was, *before.* Every object weighted with terror and history. *And then one day she was simply gone, vanished without a trace, leaving only these few inscrutable clues. Why did she leave her brush, but take her bedding? The fridge is almost empty—did she pack food for her journey? Examinations of the blank legal pad left on the desk have revealed nothing.*

You're not allowed to enter rooms like this, but then why was the door open? It swung inward with a shove, sticky with the humidity. And inside it was stifling. How could Sylvia live in a room without windows that opened? I opened the hallway door and turned on the fan which made me want to stick my fingers into it, just because the harm it could cause seemed so inevitable. I circled on the braided rug. The room felt so small, like a dollhouse room, I felt like a giant in it. There was nothing to be done here. But surely Sylvia would come back, some time. But I couldn't wait forever. And where had she gone? Maybe just to do her laundry. But the air of the room smelled so old and stale, I was sure she hadn't been here for some time.

Something else bothered me, some change in the room I couldn't identify, and I found it when I went to empty an ashtray into the wastebasket. The picture was in there, the "Last Supper" print, frame and all, the glass cracked as if it had been dropped or thrown. I looked at the blank spot on the wall where it had hung over the hot plate. The paint was too old, Sylvia's tenure here too short for it to have left any trace.

I sat down on the bare mattress and lit a cigarette. I ran my hand over the ticking, played with the buttons. Something about it drew my attention. The mattress was old, it had those ancient-looking stains that every old mattress has. But it wasn't the stains, it was the mattress itself, and slowly the reason pushed itself into consciousness, a Proustian intimation (as if Proust were the only one who ever remembered anything): I used to have a teddy bear made of mattress ticking like this,with these blue quilting buttons for eyes and mouth. Well, I wouldn't write a novel about it. His name was Mr, Mr . . . Mr Somebody. He had been my most prized possession and companion, and of course I had no idea what had become of him. Why should I care? But now I was susceptible, and I found myself patting the mattress, thinking, Oh, Mr Somebody, where is she? Where has she gone? Mr Somebody just lay there, lumpy and forlorn. He was no comfort or help at all. I yanked off one of the buttons and tossed it toward the metal wastebasket, and it clanged off the rim.

I got up and turned off the haranguing fan, and as I did I saw the blank legal pad on the desk. I found a pencil and took it and the pad and sat on the bed, and I tried the old trick of rubbing the lead on the blank page to reveal what had been written on the page above. I rubbed and rubbed and nothing emerged.

I turned sharply with a sudden clutch of fear and hope as someone appeared at the door. A female face peered in. Not Sylvia. This woman had on a tank top and gym shorts, and a towel wrapped around her hair, and another towel draped around her neck. She looked tentatively but nosily in through the door. I looked back at her, and set the pad aside, casually. She edged into the door, and raised one hand at the waist, a shy little wave, and said,

—Hi.

—Hi, I said.

—My name's Joyce, she said.

—I'm Bryce.

—Hi, hi Bryce, she said nodding and smiling. —I just moved in.

—Welcome, I said with a wave.

She grew bold, and leaned against the doorframe, smiling broadly. But then some dim doubt beclouded her countenance.

—I thought, she said, motioning with a circular wrist movement, —I thought this house was all girls?

—Oh, it is, I said.

—Oh, she said; and again the same circular motion which this time indicated the smoke in the air and my cigarette, as well as the list of rules on the door: —I thought it was, like, no-smoking?

—Oh, it is, but I don't live here, I said.

—Oh, she said. Okay. I just moved in. I don't think that Mrs Stimic really likes me. And there are so many rules, you know? I couldn't believe she was, like, for real?

—Simic. She is, though, I said. She is for real. She's very protective.

—But, like, you can have guys in here, right?

—I've never seen any guys, I said.

—But, like, *you're* a guy? she said, pointing.

—Right, I said. That's right.

—Right. Well, if you can smoke, can I bum one?

—Well, you can't, I said, pointing at the list. —But, sure you can.

She came in and took the cigarette I offered, and turned a little circle on the rug, then saw the desk chair and pulled it over. I had just finished my cigarette, but I lit another to smoke with her. We put the ashtray on the floor between us. She leaned forward to the lighter I held lit, and drew in deeply.

—God, I haven't had a cig in ages. I don't know why I moved into a no-smoking house. You're like the first *person*

I've seen here. It's kind of creepy, you know. And that Mrs Civic.

I don't know why finding me in this place where I clearly wasn't supposed to be was comforting to her. You could see how some people got themselves into trouble. I explained the situation that she didn't seem to be much troubled about. I asked if she had seen Sylvia, or Sylvia's car, or any sign of Sylvia. She hadn't. She had been here only two days. But since she'd obviously been nosing around, I was led to think that Sylvia hadn't been here. Then where was she? Mr Somebody was still keeping mum.

Then as we sat and smoked, and this girl rambled a bit—she'd just come to town for school, came early 'cause rooms around the U went fast this time of year, no one around, creepy Mrs Stivik—I imagined Sylvia coming into the room while we two sat smoking, and how that probably wouldn't help things a bit. Then the question was not where was Sylvia, but what the hell was I doing here? Even if Sylvia were to come in and find me here alone, she might be angry, and with every right, that I just walked in here as if I owned the place, owned her, as if I were entitled to anything.

I was entitled to nothing. How could I believe that? If I were to believe that, Sylvia would have to make me believe it. This much would be required of her. For though no vows had been made, still bonds were created through nothing spoken, and if you tried to escape them, pretend they did not exist, then they would haunt you. This was something allied to the secret desire, a weft in the woven energies on which one's life is arrayed. To how many people in one's life would one say: I love you. I stroked the mattress and remembered our passionate making-up, the last time I was here. Some trace of those moments remained, and compounded with the memory of Mr Somebody, it created a weird, incestuous feeling.

Joyce leaned forward to stub out her cigarette.

—God, sorry I've been like rambling, she said. I didn't mean to talk so much.

I shrugged it off. I hadn't been listening. She moved toward the door.

—Just, you're the first person I've seen here. Thanks for the cig. I guess I'll go dry my hair. I hope you find your friend. Sybil, is it? Sylvia. Right. Okay. See ya?

She went away barefoot down the hall. She walked heavily, as many women do, although she was not large. In the web of my secret desire I caught some small invitation in the way she said she was going to dry her hair, that this woman wanted to break all the rules at once and have it done with, and that she wouldn't stay here long. And I told myself that was nuts, and that it wasn't. And I thought that a lot of times it was better *not* to notice things, that maybe it was living too intuitively that drove people crazy, unless you were willing to believe and act upon anything you intuited.

How had Sylvia lasted so long in this house of stupid rules? I had never thought of her as a scofflaw. But she was. She was a subversive in all she did.

I shut the hallway door. I carefully replaced the desk chair in its previous artificial position. I tore off the top sheet of the legal pad and replaced the pad where I had found it. Crouching in the door I turned back and said aloud,

—So long, Mr Somebody, so long.

From the rooming house I drove to Carla's apartment, just across the river. I hadn't seen her since she had cooked dinner for me at the restaurant. She came down to answer the buzzer in shorts and her *Scarlet Letter* T-shirt, a red bandanna holding back her hair. She looked hot and tired. She opened the door and she wasn't smiling. She said,

—Bryce, long time.

—Hi, I said. Long time.

She leaned with her hands on the doorjambs. She was barefoot and she seemed smaller than I remembered, and her expression as she looked up at me was almost a glare.

—What's up? she said.

—Just stopped by to say hi. Are you busy?

She hesitated, then said,

—I guess not.

And she turned from the door and walked toward the stairs with nothing like an invitation. I barely caught the door before it shut, and followed her up the stairs.

In her apartment she continued walking away from me. She went into the kitchen saying,

—I think I have some iced tea.

—Great, I said. It's hot.

—Yeah, it's hot.

Then she came out with two glasses of tea, and said,

—But I don't have any ice, and I don't have any lemon.

—Doesn't matter, I said.

She sprawled out on her couch, and I took a chair. We sipped our tea. I lit a cigarette and offered her one which she declined. She looked up at the ceiling. She scratched her neck.

—So what's going on? she said.

—Nothing, I said. Not much. Have you seen Sylvia lately?

—No.

—Have you talked to her?

—No.

—I can't seem to find her, I said. I thought maybe you had talked to her. She hasn't been going to work and I don't think she's been at her room. Any idea where she might be?

Now Carla looked at me, hard. She said,

—No idea. She's not here. You can search if you want. I'm not trying to steal your girlfriend, Bryce.

I sat back in the chair.

—I don't know if I have a girlfriend, Carla.

Then we sat without talking for a minute, and I said,

—What's the matter?

—Nothing, she said. It's hot. Work is shit. You don't call, Sylvia doesn't call. Nothing. After that night, not a word. Nothing.

—I'm sorry, I said. I meant to call. Things have been really weird, Sylvia and I had a fight, sort of, and now she's disappeared.

—Did you tell her what you did? I told you not to.

—I didn't, not exactly. And that wasn't it. I don't know what it was.

—She disappeared and you thought she was here, you thought I was hiding her from you.

—No. I really didn't. It's been days since I've seen her. That other time, I thought she was here, and she was. I didn't think you were hiding her. I thought she would want to be here.

Carla leaned toward me and held out her hands as if she wanted to grab me, but doubted she could reach me.

—Bryce, I feel silly to always be explaining Sylvia to you, because I've always thought you're a lot smarter than I am. But you've got these raging hormones, and whatever other male stuff, so here it is: It's not a contest. Sylvia doesn't want just me or just you. She needs us both. She needs *people,* do you get it? As it happens, you are her lover, and I am not. But

I'm not going to abandon either of you because of that. I'm not going to stab either of you in the back. What good would that do me? So, relax. But—if she had to pick one, a lover or a friend, right now she would pick a friend. In the abstract, I mean.

She sat back and blew up a hard breath, and her hands fell *fwump* on the couch cushions and raised a little dust. I didn't feel smarter than anyone.

—I guess I knew that, I said. It's like you said, hormones, or, I don't know, just plain fear.

We looked at each other.

—We've been friends a pretty long time, Carla said.

—I know. I'm glad.

—Well don't forget.

—I won't, I haven't.

I stood up.

—It isn't easy, she said.

—No, it isn't.

—I'll tell you if I hear from her, unless she tells me not to.

—Okay, I said. Thanks for the tea.

I kept calling the lab. I had Sylvia's boss Bob good and irritated with me.

—I've left a message here: "Sylvia, your boyfriend called one hundred times." Okay?

Bob became in my mind the anti-Boscovitch, an agent of ignorance and negation. I pictured him deep down in the earth, practicing sinister arts, killing small rodents for pleasure.

But Boscovitch himself said you couldn't attain your desire. That dark sea of the soul's separation. But Boscovitch was on the side of the soul.

. . .

I looked up Sylvia's mother's address and drove out there. People who grow up in the suburbs sometimes come into the city, but if you grow up in the city, why would you ever go to the suburbs? So that open green land of ramblers was strange to me, but I didn't know quite how out of my element I would feel until I pulled into the smooth black driveway of the cul-de-sac home. It was late afternoon and everything was still, and there was not a soul in sight. The house was big, sort of modern, kind of cedary. I walked uncertainly to the front door, decorated with a grapevine wreath, and rang the bell, and heard a deep, three-tone tolling within. My heartbeat was rabbity. I wasn't sure this was even the right house, the right street. Out here they named the streets with a virtuoso disregard for clarity. They had Paradise Court, Paradise Circle, Paradise Lane, Alley, Drive, Way.

I heard fast footsteps running up stairs. The door swung open, and there stood Alison, breathing hard, eating something. She looked at me, and blinked, and said,

—Hi!

She chewed.

—Hi, I said.

She stuck her head out the door, looked down the driveway, across the yard.

—Where's Sylvia? she said.

—I don't know.

—Oh. What are you doing here?

—I just . . . I don't know, I said. Nothing, really.

—Oh, she said, putting her hand to her mouth and depositing the rest of whatever it was she was eating. It had a sweet fruity smell. —Want to play ping-pong?

—Okay. Sure. Why not?

And indeed, why not. She turned from the door and bounded down the steps to the basement, to a classic suburban rec room. I don't know why discovering the purely typical always surprises me. I always imagined that stereotypes were created through an averaging process, like average daily temperatures, where it's rare that any day's actual weather exactly fits the norm. But here was the perfect paneled rec room, with a linoleum floor and a bar in one corner, and plaid-covered furniture (*Herculon* popped into my head; *Barca-Lounger*), and a big color TV turned on to a game show. There were several watercolor landscapes on the walls, far better than the starving-artist-sofa-sized-oils that should have completed the decor. I mentioned them, and Alison said Richard had painted them. And I was given pause.

The ping-pong table sat in another, unfinished room, through a set of swinging saloon doors. There was just room for it between the furnace and the concrete-block wall. A metal duct swept hazardously close to the table at that end.

I hadn't played ping-pong since college, when Frank and I had played endless games in the dorm. I had thought I'd become pretty good, but really it was just that Frank and I had become pretty good at playing together. We had long, dramatic rallies, and almost every game went into extra points. And we played hundreds, maybe thousands of games, hours, days, weeks, months of our precious youth spent at the ping-pong table. But even a mediocre player could beat either one of us.

Still, I thought I would beat Alison easily. I should have remembered the Fourth of July badminton game. If I had, I would never have agreed to play her.

We warmed up a bit. She was proficient, if slightly clumsy. After five minutes she said,

—I'll serve, okay?

—No way, I said. Volley for serve.

She bounced the ball to me, and I tapped it back, and she hauled off and slammed it. It bounced high and clanged off the furnace duct.

—What are you doing? I said. It has to go over three times.

—That was three. Me to you, you back, I put it away.

—It has to be *hit* three times. You *threw* the first one.

We argued, and finally did it over. She won, and served the first ball into the net.

—I get to do that over. Starting the game it's first ball in.

It didn't improve from there. She produced endless arcane and questionable rules about where the serve must land, touching or bumping the table, interference by the furnace, the hanging light. I don't know why even a simple game of ping-pong had to be so difficult. I became more and more annoyed as Alison pressed her legalistic chicanery, scored points, won the first game and then the second. I won two, but narrowly, and I wanted to cream her. I tried to stifle that sentiment. Lately I was irritated by everything, I reacted contentiously to everyone. But also, Alison was being a brat. She won a game, and I won a game, and she won the last one for four out of seven.

—That's it, I said, putting down the paddle. —You're the champ.

—Let's play again.

—No thanks.

—Just three out of five.

—No.

—Two out of three.

—I'm done. You win.

As I pushed through the swinging doors to the rec room the front door opened, and I heard Irene call,

—Alison? Are you home? Whose car is that out front?

I came to the bottom of the stairs, feeling as if I'd been caught at something, thinking that breaking written rules was easier than trespassing where you couldn't see the bounds. Irene stood at the top of the stairs, and when she saw me she gave a surprised smile.

—Bryce! I thought, when I saw that *car* . . . But I didn't think . . . Where's Sylvia?

Alison had been brilliantly oblivious to the strangeness of my being here alone, but Irene, I knew, would never believe that I had just come out for a game of ping-pong. So I started up the stairs, saying,

—She's not here. I don't know where she is. I thought you might know.

—I don't, she said. You know, she doesn't call that much.

She looked down at me as I looked up. She had just come from work, had a briefcase in her hand, looked hot and tired, like everyone else in these dog days. She saw me, and answered what I didn't even know I was asking:

—Let me get changed, then have a glass of wine with me.

Nothing begrudging in it. Her eyes, however, mirrored the forlornness that had been rising in me and was ready to spill over. I liked her at that moment as much as I ever could. Maybe she saw me as a way toward her daughter, who was far away from us both. Did it matter? Sooner or later, you have to accept the commerce of human relations. Whether it is blatant or subtle, it is always there.

Alison came out of the ping-pong room, slinging the swinging doors extravagantly behind her.

—I'm so bored, she said. How about croquet?

—No, no more games, I said; but I smiled inside as I said it.

Irene changed from her work clothes and we took our glasses of white wine onto the backyard terrace. Alison hung around for a while, until the tedium of adult company overwhelmed her, and she left, with muttering to let me know I had betrayed her. Irene asked about the car, joked that I had probably driven it more in the months I'd had it than it had traveled in the twenty years previous. Then she said:

—I liked you the first time I met you, Bryce. I thought, Sylvia has finally found a good one.

Then we were being honest. I said:

—I know how to make a good first impression. It doesn't really take that much. I'm not overtly obnoxious, right away.

—That's not it. Don't sell yourself short.

—I don't think I do, I said.

—I'm pretty good at reading people. I'm almost a professional, remember?

—Right, I said. Psychology.

—I think you're good for her, Bryce. You seem steady, she needs that. She's just always so anti anti anti. Against everything. Everything I do. Everything I say. She seems to blame me for everything.

—I don't think she blames you, I said, not wanting to get into this, not wanting to be between this Scylla and Charybdis of mother and daughter, regardless of what I might learn, classes were over, it was time to show what I knew.

—It's not uncommon, I know, Irene said. Blaming the mother is the national pastime. Parents make mistakes. They're people too.

I wanted to ask: then you explain it (I didn't want an explanation, explaining wasn't—); whom do you blame? She was studying psychology, she ought to have a clue. But I couldn't blame her for being a clueless expert, because I didn't believe in experts. I believed that any given discipline, any chosen course in life, was no different from deciding to drive circles on the one-way streets or not to cross the river.

—I really don't think it's a question of blame, I said.

—No, you're right. It's an old, old story. Do you get along with your parents?

—My father is dead. I get along with my mother, I said.

—I'm sorry, about your father. Sylvia didn't tell me. Sons and mothers, and fathers and daughters, they usually do all right. Sylvia seems to get along with her dad. It seems trite, though, doesn't it? It's all that Oedipus and Electra crap. You think that something so obvious can't be true.

—I like your mother, Sylvia's grandmother, I said.

Irene laughed.

—She's something. She has that pioneer spirit. The last of a dying breed.

This brought a pause. There was no mention of her father. There will be no mention of her father, there will be no further mention of my father. The dead fathers are better forgotten. But then Irene said:

—Sylvia makes things up, you know. Or anyway she exaggerates things. I think she does it to make reasons to pull away from me. That's the hardest part about straightening things out in this family: we can't even agree on what actually happened. Sylvia has always been dramatic, and she was alone a lot as a child, so her imagination tends to run away with her. When you say things about people who are dead, who can't defend themselves, that's just not right. Sometimes you have to let sleeping dogs lie.

That seemed an odd doctrine for a budding psychologist, but I was willing to allow her the contradiction. And I countered the uneasiness of gossiping about Sylvia with her mother by telling myself that it might all be academic. Sylvia might never reappear in my life. Half an hour ago, that would have been unthinkable. Now it was convenient and effective. No more games. We were being honest. Indeed. I have mentioned my belief in my inherent sincerity, pessimistic and ironic though it might be. Now I was beginning to think that sincerity was simply the most disastrous approach one could take to life, to love, to anything. I thought I might be able to be more optimistic, less ironic, if I could manage to be less sincere, not say what I was thinking, not even think what I was thinking, stop giving straight answers.

—Her father has tried to turn himself into some kind of saint, Irene said. But that doesn't change anything that happened, it doesn't right the wrongs he did. I haven't changed the way he did. So maybe I remind her more of those bad times.

Another dead father, her ex-husband murdered by self-reform.

—I really don't know what Sylvia thinks about all this, I said. I haven't known her that long. It seems long, though it has gone fast. I've never met her father. Obviously there are problems. I don't think children should always blame their parents. But I don't think they necessarily shouldn't blame them, either.

It required a moment for both of us to sort this out. Irene said:

—Do children ever blame themselves for what's wrong with them?

—Do you blame yourself?

—For what?

—For what's wrong with you. I mean, for problems in your life.

—Yes, of course.

—But maybe you shouldn't. Don't you think Sylvia blames herself?

—I don't know what Sylvia thinks, truly. Sometimes I can guess, most of the time, no.

—But you know how strongly she's tied to you, to her family.

—She used to tell her teachers that she was an orphan, that her parents were killed in the war.

—I think that just proves my point, I said. The ties aren't always comfortable. We look at other people and we don't have a clue what makes them tick, what they think, what they want. But we think everyone else can look through us like a glass of water. How odd.

—You're right, it is strange. It's a serious contradiction.

—I've begun to think that there's one unique impulse that drives each of us, and everything we do is somehow in its service. Have you come across anything like that in your psychology texts?

—Well, sex. Survival.

—No, it's not that blunt. It's unique to each person. It's as complex as your genetic make-up.

—I'm not sure I know what you mean.

—I'm not sure I do either.

I was feeling energized, as if something was about to begin. I wasn't being deliberately insincere, but I was beginning to play the game, beginning to understand the rules. I said:

—If you can start to draw even the vaguest picture of this impulse, this drive—I've been thinking of it as the secret desire—you'll be way ahead of most people. But it's not a contest, I don't mean it like that. You'll be ahead of yourself, ahead of all the stuff that drags you down. Like this worrying about who's to blame, and what really happened, and all that. That stuff will stop mattering, and you'll be able to move forward.

—That's the aim of therapy, Irene said.

—No, it's not, not really, not in the same way.

—I don't really know what you mean, then, she said, and she seemed annoyed, and I didn't mind.

I had finished my wine. I was ready to go. I asked Irene if she had any idea where Sylvia could be. She didn't think she would be at her grandmother's. She said Sylvia's father was out of town, and she might be at his apartment. At the door she said to me,

—I'm not trying to poison you against Sylvia, Bryce. You know that, don't you? I really do think you're good for her, and I only want what's best for her.

—I know you do, I said. We all do, we all want what's best.

And I meant it.

Tantalos

We live in memory and by memory, and our spiritual life is at bottom merely the effort of our memory to persist, to transform itself into hope, the effort of our past to transform itself into future.

Miguel de Unanumo, *The Tragic Sense of Life.* She discovers the passage, underlined, in her father's book. She sits cross-legged on the shag-carpeted floor of her father's apartment, with several dozen books pulled off the shelves and flung widely about her. She smiles when she sees that her father, whose sense of irony is somewhat stunted, has failed to mark the brilliant finish to the soaring passage:

All this, I know well, is sheer platitude.

Oh, Dad. Don't you get it? She doesn't feel superior, exactly, but bemused, and intrigued. Never in a million years would her father think to note the bravado disclaimer, which sent the passage every way at once. Platitude, but what grand platitude. What *sheer* platitude. Still, platitude. If her father were an editor he would have deleted that kicker without qualm. It was a generational difference, she thinks. She doesn't feel, really, a part of any generation at all, but would you, if your generation was defined by irony? To look ironically at an ironic generation, where would that lead you? That was why her father had no patience with irony, no capacity for it. Irony is the last refuge of the ambivalent.

Her father was confused, to be sure, but not ambivalent. He was off again to the Islands. Some blow-up with his girlfriend had sent him off alone, though the trip had been planned for months. It always seemed to happen that way, as if he planned or anticipated personal crises and vacations in tandem. He asked Sylvia to water his plants and bring in his mail, and on her first visit she decided to stay. It was a kind of musical chairs of the lovelorn: her father hopped on down to the Islands, and she slid into his place—but who would fill her small room? No one could. Why would anyone try?

But then why would anyone want to try to fill her father's place, to live in his condominium hall of ghosts? This was the purgatory he chose for his perdition, after the divorce. But you must be careful about real estate decisions based on moral shame and not on market realities. This place was an albatross, he often said. The assessments and fees kept going up, but he would never be able to sell the place. A "garden level" apartment in a ticky-tacky complex, a pioneering development to provide the satisfaction of home-ownership to those who couldn't afford or didn't want a house. But now there were thousands of cheaper, nicer, better-located apartments all over the suburbs. Whenever Sylvia came here she was freshly awed by the extravagant awfulness of the place. It was a sprawling, accidental-looking development wedged between obsolete highways, with numerous, tumorous wings, els and adjuncts, the whole thing dressed up in fake tudor which had weathered badly, the cheap brown stain of the warped decorative boards bleeding cloacally down the deteriorating stucco. A few twiggy, ravaged saplings adorned the common green, only nominally green, a sea of mud in spring and autumn, birthplace of dust storms in summer, and somehow dingy even under snow.

The tenants came mainly from two distinct and disparate groups—young working-class yahoos and retirees. It was fascinating, sociologically, that you could achieve such a supremely volatile mix by sheer chance. The retirees complained about the yahoos' cars and parties. The yahoos retaliated by going after the retirees' pets. There was a thirty-pound limit on dogs, and someone noticed that Mrs Veske's portly cockapoo had been growing ever portlier. An official weighing was demanded. Muffin tipped the scales at thirty-four, and was banished. The retirees came back with demands for enforcement of the stated-occupancy limits, and whole hives of yahoo squatters and sub-letters were flushed out. And on it went.

Her father's enormous guilt was the sole cause of the desperation and fear the place inspired; Sylvia was convinced of this. She could never remember its name; it blended with those of all its tacky-idyllic ilk: Valleyvale, Glenglade, Creekbrook, Vistaview.

Her father's apartment faced south, but here at garden level—three-quarters basement—grade was at eye level, and the building across the green blocked most of the sun. On the outside walls faint patches of moisture developed like your most ominous dreams. The ceiling was low, even less than eight feet, Sylvia suspected, and when she was here with her father she felt the urge to cry out, every time he stood: *Dad! watch your head!* In lieu of baseboards, the shag carpet traveled six inches up the walls. The corners of living room and bedroom were lost in shadow, no matter how many watts were set blazing. The kitchen was comfortable, and fairly bright, with all the lights on.

The desolation suits her, since she has come here as a refugee. With the heavy drapes drawn, with nothing she must do, she doesn't bother to determine day and night. In three days she hasn't left the apartment or even looked outside. She scavenges food, cooking nothing, eating nuts and berries, literally, and helping herself to the alcohol that her father heriocally, insistently serves to guests while he himself abstains. She mixes strange cocktails, and when she has drunk enough of them imagines that she will gain fame as a revolutionary mixologist. Vodka, carrot juice, and sweet vermouth—the Bolshevik Sunset is born.

She watches television through the night, old sit-coms and recycled made-for-TV movies. They make the world seem frantic and desperate. They give the impression that no one knows what on earth is going on here. Sylvia learns that every time someone leaves or enters a room, you are supposed to laugh. She wouldn't mind laughing, if only they weren't so blatant with their jokes. It is an enormous industry of gallows humor; everything they don't mention, all the real terrors of the world, press and seep like water through the creaking, leaky boards of their unfit ship. It seems this way when you watch five hours straight, drinking strange cocktails and eating stale healthy crackers.

We live in memory and by memory. . . .

Her father lives in a museum. Wandering slowly through

the apartment, unaware of night or day or any world without, Sylvia begins to see what she has been afraid to see: her father's monumental effort to make memory persist. It is the single, essential theme of his life. It is the ruling motif in his interior decorating. But you can't call it decorating because there is nothing decorative about it. Things have been put up in certain patterns, there is symmetry here and there. But what has been put up, and there is so much of it, has everything to do with remembering, and nothing to do with aesthetics. It is anti-aesthetic. It is a statement of anti-taste.

Photographs, to begin with. Dozens (at least: Sylvia cannot bring herself to consider that there might be hundreds) of pictures of her and her sisters, at all different ages, in all different groupings, in all different uniforms, engaged in all different activities. These are hung in cheap drugstore frames—even at that, it has cost him hundreds of dollars. There are a few old black-and-white and sepia-tinted photos of his parents and grandparents, these grouped together under a many-windowed mat and framed in gray barnwood from the family homestead. A crooked old farm implement hangs on the wall nearby; Sylvia has no guess what its function was. The photographs are most densely clustered on the living room walls, and spill down the hallway, into the bedroom, the bathroom. Only the white walls of the kitchen's eating area are relatively free of them. These walls are filled with nautical maps and paraphernalia.

Along with the photographs there are other artifacts of Sylvia's and her sisters' childhood: a poem about sailing that Sylvia wrote in junior high, which her father copied out in florid calligraphy and framed; someone's little handprint pressed in a circle of clay, set off with a glossy yellow finish; certificates from spelling bees and science fairs; three pairs of baby shoes suspended in a net that hangs in one corner like an enormous cobweb (it also holds shells, sand dollars, colorful glass buoys);

a spoon and fork that Sylvia carved out of driftwood, idly, one summer at Lake Superior; a Father's Day certificate awarded to The World's Best Dad. Even grade school drawings he has saved and framed. One of Alison's shows a house in the foreground, with a man and a woman and three little girls beside it, and above to the right, just below the yellow sun, a tiny flat-roofed house, and next to it, tipped in the sky, a big man with glasses and a beard, much larger than any of the other figures, and this is Dad.

Sylvia never notices these things. They are the white noise of her father's memory. She begins to look, to see, and then she must stop. She tells herself she isn't shirking: she gets the idea; she doesn't have to flail herself with it.

. . . *our spiritual life is at bottom* . . .

If you don't want to think about things like this, if you don't want to think about anything, then you can always watch TV. For most of the first three days, Sylvia has watched TV. Sitting in her father's enormous reclining chair with the remote control box in her hand, she calls the television to life and quickly descends into a video daze. Since she has no TV in her room she has not built up immunity, and so her father's big color set is mesmerizing. She doesn't watch TV so much as operate it. Her father rarely watches television, but he has cable and all the movie channels, for when the girls—mainly Alison—come to visit. Cable is something Sylvia associates with dingy condos, she thinks of it as inherently tacky and depressing. Cable TV makes her imagine empty, grimy, shag-carpeted apartments, with the black cable cord emerging kinked from the wall like the tail of some animal that has not quite escaped. The kind of place where, in a smaller, southern city, a loser's desperate last stand would unfold, and end in ignominious death, blood on the shag. These are the apartments that people take when they have no other choice. Cable TV is an inducement. It makes prospective tenants say: Well we've come to this place where we live without hope, the future shows only waste and despair, and in our better days we might have made a pact to kill each other if we had no choice but to live in a place like this, but—*It has cable!* We'll take it.

There is really nothing better to watch on cable, unless you like arcane sports, but it makes it more fun to operate the TV. Sylvia goes through the channels a few times, *moderato,* pausing with each to understand what's going on. On sixty-three channels there's nothing she wants to watch, but she keeps going around. Gradually she builds speed, and she plays a game with herself, trying to identify the program by its split-second appearance as she jabs the button and races through. *—ter Gr—*; *—is Jim R—*. She gets these two old favorites from the rerun channels,but since she hasn't watched TV in years she has little hope with most of them. There are so many attractive, sarcastic families, adolescent parents and precocious kids. The teenage boys are appalling: smug, officious, but presented as charming. They're going to rule the world. But isn't it more likely that they'll wind up like ineffectual Dad? Have the writers considered heredity? We want to be free of heredity, masters of our own cloying destiny. Anyway, everyone has a good time. Not that hard, when all you need to do to get a laugh is leave.

There are a few pay channels that her father hasn't picked up yet, and these appear as dark, wavering color-negative images, like infrared photographs. Sylvia pauses on these, from time to time, and tries to decipher the action—it is a more active pastime than ordinary viewing. The image rolls and flickers and jerks; thick black lines shoot across; the characters approach definition, then dissolve. The only sound is a staticky fuzz. When the picture holds steady for a few seconds at a time there is a great urgency in the actors' efforts to communicate to each other, to Sylvia, desperation as their strange mouths move with comic speed, soundless, before they are swept away again by the electronic whirlwind, and whatever they meant to say will never be heard—the last garbled message from the sinking ship, the arctic outpost, doomed voyagers to the sun.

The predominance of '70s sitcoms and old movies tells Sylvia that it is night. It could be daytime on an obscure channel, but there is also the deepening feeling and the sense of a dormant world. It makes her feel quiet, it presses the need to tiptoe, to whisper if she speaks, to sit quite still. She is somehow aware of the others, all those others who watch TV through the hours that weigh like cold gray steel. Night is still night, and always will be. She feels it, and even in her stillness, it makes her feel wild—a little twitch behind her eyes, a thin vibration in her chest. She never knows the precise time—she has taken off her watch, and the only clock in the apartment, a digital clock radio in the bedroom, blinks 12:00—12:00—12:00 because of a brief power outage. Perpetual midnight, the twelfth hour, the symbolic brink. The blinking brink. No time.

Then she sleeps, a heavy dreamless sleep, a sleep like tons of warm salt water, deep gray but shot golden, she swims in it, her sleep is a slow heavy swimming. And she wakes slow. Moves slowly around the apartment in underwear and a T-shirt, or sometimes in nothing at all, with a cup of decaffeinated coffee. Her skin is warm and moist. She looks at the pictures, though she doesn't mean to. She doesn't know what she is doing, she isn't thinking, a rarity for her who often thinks even in her sleep, solves problems in dreams. She merely takes in the pictures.

. . .

. . . *to transform itself into hope* . . .

Does it? She thinks not. That is the catch: simply remembering isn't enough. Surrounding oneself with one's past is not enough. Gather it up and drive it forward. This is what is required. We have two things that we call "our lives": what was, in memory, all that we have done and that has happened to us, all we have thought and felt; what is, the current circumstances. How do we bring them together, roll them up the mountain of memory, shove them snowballing down the slope of hope, of to be? Dad can't tell you; Sylvia can't tell herself.

Sitting in her father's raging clutter, Sylvia thinks of her own bare room. She sees it from the outside. She sees a face: the windows flanking the door are funny vertical eyes, and the door a big nose, and the fire escape landing, viewed from the side a black line, is a tight-lipped mouth. It sees all and tells nothing. It's the deadly serious face of a clown. It is neither happy nor sad. It stands to be mocked—*Look at that huge square honker!*—but it is unmoved.

The barrenness inside seems so opposed to her father's rooms, but Sylvia thinks they are on the same quest. They are following different routes, by surfeit here, by negation there. Preoccupation with the past, with the life that was, is their common theme. But is it a quest? Is it a quest: are they moving forward, or running in place, or just standing there, the only motion a slow erosion?

On the third morning—or say, the third waking—she stops in front of a picture: Sylvia in the Islands. Here she stands against the aluminum mast of a sailboat, under blue sky, over bluer sea. She wears a green bikini. Her back is arched against the mast and her hands reach behind to hold it. One knee is bent and extended, a kind of languorous stretch. Her head is lifted to the sun, her eyes closed. Her hair is in the Zelda bob, and strands blow across her closed eyelids. Her skin is rosy, glowing. Look at this throat, chest, belly, these shoulders, breasts, thighs, hips. All is roundness, fullness, smoothness. It was the height of the Zelda time; her health was probably terrible, but this is the first full burst of womanhood, and she has never looked better. The photo is so boldly sexual, it embarrasses Sylvia, who stands naked, looking at herself in the picture. What was her father thinking, putting that on the wall? Then a slow longing begins in her, she wants the girl in the picture, she wants to touch, and be touched as the girl in the picture. She almost cries out with the desire that grows and grows, but slowly, still slowly, and going on slowly. She takes a long hot shower, and soaps herself repeatedly.

She remembers how sexy the world once seemed. She has come to know again what a power sex is, that seems to come from nowhere. Its intimations went as far back as Hermes with his little cold bronze balls and cock, a dark mystery. And the horses when she rode. And the summer on the farm: the texture and scent of earth; the silk that burst from swelling cobs; the animals unabashedly copulating whenever they were so moved; and birth, the bloody emergence of life. Why didn't anyone *talk* about it? It was all so exciting, alarming, astonishing. Did only she feel the wildness, the maddening pulse, the blood heat? No one said a word. When the dogs went at it they pretended not to see. Sylvia could not fathom grown-ups. This was the only thing worth talking about, at all. It culminated in the Zelda time, and Zelda infused it with legend and grandeur. For Sylvia no attachment has ever been trivial. It has never been just for sex. Not because sex is trivial to her, but precisely because it isn't.

Then after her fall, her return in disgrace, the world had drained of desire. And she had seen it through, that bad bad time, and in time the desire returned—not just for sex, for life itself. And the world with desire was surely more vital, if not entirely happy or serene. She thinks of the bad time, she thinks of the counselor, that big bearded Lutheran Miss Lonelyhearts, and of her own zealous despair. More than a year it had taken, he had weathered another winter with her, her third since starting college. Then she felt more stable, she thought she could manage, and the sessions became to seem like a crutch she would have to throw away if she was ever to walk on her own. The next winter had been difficult, she was still very much alone, and far from happy. Then the spring, and the lover. And everything was okay. What had happened to nihilist Sylvia, evangelist of despair? Was she just another lovelorn case, after all?

Very clean, and somewhat soothed, and dressed, she moves about the apartment with a slight restlessness. Till now she has been content with her vacant occupations, an end-of-the-semester, last-exam feeling, in which nothing was exactly what she needed to do, nothing what she needed to think. A ray of sun penetrates the dark curtains, and she realizes that it has been raining since she arrived. She remembers, as if it were weeks ago, that she arrived under the gray rain, feeling like a criminal on the lam. She came in with her sack of loot—just her laundry. But yes, a crime has been committed; yes, she is guilty. What crime? No matter. Guilt falls like rain.

The restlessness is building, and the TV won't divert her. She turns it on and discovers a channel with fish. That's all—

a steady shot of an aquarium, tropical fish gliding among plants real and fake, through little castles and windmills and shipwrecks. There is music, orchestral versions of Top 40 hits. She turns off the sound, she leaves the fish. She goes to the bookshelves and peruses the weirdly heterogeneous collection. The books are treatises on navigation and knots, mathematics texts, books of puzzles and games, children's books, Meister Eckhart cheek to cheek with Kahlil Gibran; Walt Whitman rubbing elbows with Hallmark; Robert Service, Robert Frost, and *Modern Accounting Techniques;* popular novels, serious novels, *Legumes and You.* One whole bookcase is a yellow wall of *National Geographic* magazines. It is the *Geographics* alone that make her feel depressed and uneasy. Because this same collection squats ponderously in thousands upon thousands of middle-American homes? Because you only ever look at the pictures and never read the stories, and never really learn anything? Because the glossy stock, the technically stunning photographs, the self-important, lofty air of the whole endeavor make the world and its peoples seem at once strange, distant, stilted, boring, theatrical, and incomprehensible? We've been to places you didn't know existed. We've seen stuff you couldn't even imagine.

Because it's like a zoo, Sylvia decides. Because they cage up real wonder and smother it with the sleazy garb of education and science. We just want to see the weird animals. We want to see the bears beg for popcorn.

She pulls off books at random and seats herself among them. Opening *The Odyssey* she finds Odysseus in Hades, talking to ghosts:

> *But he gave no reply, and turned away,*
> *following other ghosts toward Erebos.*
> *Who knows if in that darkness he might still*
> *Have spoken, and I answered?*

The warrior Aias, son of Télamon, the great shade burning still with bitter envy and pride, even in death. Then here is Tantalos, whose thirst is beyond mortal knowledge, who will never drink again— . . . *he burned / to slake his dry weasand with drink*. . . . Dipping his chin to the cool pond's surface, he sees black mud baking in a wind from hell. Sylvia sends a thought to her father: *Tantalos, c'est toi.* And then Sisyphos, dear Sisyphos. And Herakles reclining with Hebe, her "of the ravishing pale ankles." That was good. Herakles tells of being sent down to the underworld to hunt the watchdog of the dead—no sweat, he brought the beast back up, with Hermes and gray-eyed Athena showing the way.

An enlightening visit, Odysseus has, but then all his comrades, friends, and foes must leave him, fade away down vistas of the dead.

Sylvia picks up the Unamuno. What an intriguing title. The book falls open to the passage on memory.

. . . *the effort of our past* . . .

Sylvia goes to do her laundry. She listens at the door before going into the hallway. When she opens the door she finds the hallway even darker than the apartment. She knows there's a laundry room off this hallway, but she can't remember where. She takes a chance and goes right, toward the darker, shorter end, away from the main stairway and a possible meeting with another tenant. She scurries with the overstuffed laundry bag, dragging it bumping along by the closing cord, and she almost laughs out loud with relief when she finds she's chosen right. The laundry room is empty, the metal door thumps closed with a hollow sound behind her. The laundry room is even dimmer than the hallway. She loads two big machines, and just before she pushes in the coins she hears footsteps in the hall. Coming this way? They pause. Just outside the door? Do they? She waits, holding her breath, her hand clammy on the metal coin shuttle. No sound. She waits and waits, she's frozen now. But breathing, at least. She's listening hard and hearing nothing, but the incidental sounds of the building—heating, ventilation, water—are amplified to a roar by her concentration. In this indefinite rush of sound her thoughts begin to wander, she forgets what she's trying to hear, as something tries to push itself into consciousness, some realization. She's thinking about her father's apartment, the terrible glut of memorabilia, it's something to do with those things, the pictures, the souvenirs, to do with memory—

—*slam*—

goes the coin shuttle,

—*whoosh*—

goes the water and

—*thump*—

goes Sylvia's fist on the washer lid, and thump again: It isn't to remember—it's so he doesn't have to. This was what kept her looking at the pictures she'd seen a hundred times and then ceased to see, something too obvious in the sentiment they enshrined. With his memories framed and hung on display he is absolved of the need to remember. And of course, Sylvia thinks as the washers begin to agitate, he didn't sit around maundering and wondering on the past. He is Tantalos exactly, surrounded by what he most desires and cannot have, but a Tantalos who concentrates on the twitch in his eye, the itch on his neck, who examines the foliage and criticizes the gardeners, and tells himself he isn't thirsty. Oh, if it's not inconvenient, then I might have a sip of something—but, no, no, really, don't go to any trouble; I only drink bottled water, anyway.

She goes back to the apartment, not scurrying now, not

afraid of being discovered. In the apartment now she looks again at the pictures, and she shakes her head and mutters, and wonders how she could have been so easily duped, and how she could possibly have imagined a kinship of memory with him. She curses him, but affectionately—he is good. Part of her wants to pull all the pictures off the walls, smash them up in a pile, pour vodka on the whole thing and set it ablaze. But she enjoys a game, as well, and she relishes the thought of being here with him, knowing his secret. Now she looks carefully around the apartment. She wants to understand what has happened here, she wants to get it in her head and hold it there. She knows this museum was not built casually. He considered each object, and was pierced and torn by the memory each one did truly invoke. But then it was done, and that was that. He has a mind that can do that, a mind that prefers to concentrate on the actual and distrusts anything abstract, unless it has to do with numbers, and looks hard askance at the intangible, or god forbid, the ineffable. Yet he is a devout Catholic now. Or should that adverb be *thus?* Maybe every picture and object holds a specific meaning for him, fills a niche in a system only he knows. She goes back to the photo of siren Sylvia at the mast, all rosy, round, and nubile. She looks, then looks away.

She knows she cannot prove this, how her father has manipulated memory to his ends. If she confronted him with it, if she could even arrange the feeling in her mind to speak of it, he would deny it. Of course he would; otherwise he could not go on living.

On the television the fish swim on, angelfish, glowing neon tetras, a garish purple siamese fighting fish. Tiny catfish suck along the bottom. Sylvia marches across the room and pulls open the curtains, and cringes from the sudden stark light. Unsquinting her eyes she looks out through sliding glass doors that lead from the living room to a sunken patio, a natural collection point for leaves, dirt, and litter. The wind blows strong from the northwest, birds and trash sail down the green between the buildings. From the sun Sylvia guesses it's still late morning. She hasn't been up an hour yet, but she feels tired, so she unfolds a grimy chaise lawn chair, and goes inside for paper towels to wipe it off, and comes back out with the towels, *The Collected Poems of Wallace Stevens,* the Unamuno, and Pogo. And as she lies back in the sun she is astonished by this sloth which envelops her. One load of laundry and she's wiped out. She picks up the Stevens, but forget it. She looks into Pogo, but even that is beyond her. So she just lies back with her eyes half-closed. The warm sun and air are soothing. Through the doors she can see the television, dimly. The fish go back and forth. She can't really even see fish, just motion on the screen, which fits the pattern of fish. Tropical fish. Her father could be snorkeling among them right now. She imagines her father's froggy face gurgling into the picture on the TV, the gentle angelfish fleeing in terror. Is this a man to be feared, except perhaps by fish? She knows that his heart harbors violence, she has seen it; but for years now he has contained it. Could he really be so devious in his treatment of memory? The truth she felt, she knows, cannot be proved, and its force begins to fade, even as she tries to hold on to it. Why can't she hold on to it? Most of what we believe cannot be proved. With the exception of certain trivial, mundane, and mathematical questions, nothing can be proved at all. It can be shown, for instance, that a diminishing sum will never, ever, reach a certain limit, though it draws infinitesimally near; that Mrs Veske's cockapoo is four pounds over the limit; that the driver had been drinking. But can you prove that your father is so tormented by memory that he has chosen to put it out, like putting out a cat, while still giving the appearance of being devoutly attached to every past instant of his family's life? If you took it to court, could you show beyond reasonable doubt that one's mother has ruined one's life, or that her father, through her siblings, is responsible for the scars she bears and tries to inflict on her own children? Publicly, we believe this. In our own solitude, can we be content with that belief? Sylvia thinks that she continues to see her mother precisely because she can't truly blame her. It is in the limits: the psyche is a country without borders. Sometime, you must draw a line.

Sylvia remembers being down by the river, patting the sandstone, saying: *This is the rock on which my church is built.* Where did that come from? Just popped into her head, and convinced her straight off. Then what was the cause of what followed? Why did she cry, and break, and run, flee from her very temple? She finds her left hand clenched, the fingers stroking the palm. She still feels the force of the blow, her flesh brought to his with force, meant to injure. When so recently they had brought flesh to flesh, to heal. Couldn't he see all she had put at risk?

The world is a world of things, and that is all. She wants to believe this, and doesn't want to. She shares her father's tendency, which is also her mother's, to cling to the concrete. None of them truly trusts anything they can't quantify, rap with their knuckles. Some kind of inbred skepticism laces every cell. Sylvia can dabble in mysticism and poetry precisely because she doesn't take it seriously. It doesn't present the risk it would if real commitment to it were a possibility. Yet she is drawn to it. Yet it seems dangerous. The family mythology says that Irene was visited by the Virgin Mary when she was locked in the cedar closet. She was going to be a nun. Irene is the bloodiest pragmatist of them all, yet she saw the Virgin in a closet. And her psychology study, Sylvia thinks, is a brilliant ruse, equal to her father's museum, but even more cunning. She will be an infidel in the temple, she doesn't believe in that crap one bit. What defined Irene was not suffering, or even cunning, but endurance. She would survive anything. But when survival is the highest, only good, what subtleties are lost in its blaring imperative?

Have we considered heredity? That is perhaps the worst part. That Sylvia is her mother's and her father's daughter, with all that implies, with all that came before. The tiger cannot change his stripes, and why? Because he's a tiger, stupid. But we are not beasts, are we? We have the gift of thought, and of will. We can do something or not, as we choose. We can think of something, and want to do it, and choose not to. We can be whatever we want. Right? We want to be free of heredity, makers of our own fate.

Once she thought she'd got away. In the Zelda time, and just after. How she soared then! Everything was possible. She flew to California. Then she fell. Was it the dragging family gravity that pulled her down? Then her mother didn't want her here, and didn't want her there. And then Sylvia wasn't here, and wasn't there. The nowhere time. The family gravity dragged her to the edge; and held her there, did not let her fall. It made her crazy and then kept her sane. It was, after all, the one thing she would always belong to.

Sleepy in the sun she thinks: How they go round and round: she, her mother; she, her lover; father, mother; lover, mother, father, she.

The sun on her eyes, the wind.

I did something really wrong, he said.

The lover is a medium to make a present, make memory persist, transform past into future. One must be willing to transform past. Yes. Don't care what's proved or not, don't care who's right or wrong, say: *je m'accuse.* Then roll it up Hope Hill, shove it over, down the other side, the whole pile of shit maybe rolls itself into gold.

He spoke of an idea to which she clung, from which all others were excluded. She wonders if she has an idea like that.

She watches fish. She sees the life that was as water, the life that is as stones. *Clatter, clack,* says the is; *hush,* goes the was. The was is water rising. The was is water, hushes is; hushes is, the stones, the stones that, damped and hushed, endure; is always is; was only was. Above the is, the temple stones, she swims in was, a warm narcotic weight. And all around her, what are these, these golden flashing shapes, these fish that swim between the bending liquid roof of was and these enameled stones, this is? Oh! Is that, are they, hope? They swim upstream and never reach the source; swim up and are washed down again, the fish of hope, the Sisyphos fish.

The sun is soft, warm on her face, she remembers the sun. The wind is the wind on the corn, the wind on the tall grass. Someone walks above. But here the neighbors do not greet. Sylvia says to herself: *See, here I am, I'm not hiding.* She is down in the tall grass, and someone is walking above, and she feels the sun and hears the wind. She drifts to the time when she rode her pony on the river bluffs. Gabriel is the pony's name. This time is the last time of some summer. School just a day or two away, a sick feeling already. Out on the fields in the afternoon, she feels better. She holds the reins slack, if the pony should stumble; the footing is uncertain. Big white clouds tumble by, fast on the wind. Autumn in this air, the dry wind of fall, mornings cool when she bicycles to the stables, the pony's cloud breath as she combs him. She lets the pony wander where he will, imposes no direction. Squall lines on the grass, like waves in wheat. The wind loud on her ears, and she is far away from all that threatens her, a rhythm rides her away:

then:

below to the left, a flash of color, and a louder, urgent wind: a sudden shape on wings: ring-necked pheasant, Sylvia thinks: bird explodes in color and sound, pony shies: Sylvia flies, up and up, *I'm lying on my back in the air,* she thinks. She is. Floating, then falling. Down and down, as in a dream.

She doesn't remember landing, but here she lies. *Gravity brought me down,* she thinks. She is pinned to the earth, stuck like a licked stamp. Something troubles her: she isn't breathing. She breathes: good. Breathing now, she blinks. She can blink, breathe. Otherwise she cannot move. She lies spread-eagle in the tall grass. No panic, no desire to move, though she can't. Above her she sees the pure blue sky, the big white clouds going by so fast. She hears the wind. The sun warms her face. She wonders if her back is broken, or her neck. She feels no pain. She feels the dry grass against her wrists, her neck. She lies listening to the wind, which sounds distant, and watching the clouds fly past.

She feels the earth spin. She sees the earth from a long way off, as it rolls through the darkness of space. Distinct white puffs of cloud encircle it. The earth is oddly lighted, not by the sun, there is no sun, no stars, only the great ball rolling in blackness—no, the color is deepest violet. The earth is covered in tall grass that blows in the wind. One feature only appears on its moving surface—no Atlantic, Everest, Antarctica, no Great Wall of China or Barrier Reef. One feature only: a girl-shaped shadow in the grass; Sylvia as she lies. The shadow moves across the surface as the earth spins round, it vanishes, then reappears. The earth is spinning and also rolling through space. We're going far away, she thinks. From what? From the sun? There appears no sun. From home? Maybe. It feels like far from home.

Then she hears the pony, circling her in the grass, perturbed. She wiggles her toes, and they move. So do fingers, hands, arms. She sits up, and looks around at the emptiness that surrounds her. She doesn't know how long she has lain here. If a girl falls from a pony in a field, and no one sees her fall, did she really? Gabriel is wary, still spooked, as she coaxes him to her. She remounts, and returns at a walk to the stables.

Later, she dreams this repeatedly. In the dream, everything occurs with dramatic inevitability, something set in motion, inexorable the moment she mounts. Each moment seems expected only as it arrives, a sensation more subtle than déjà vu. The pheasant appears as if on cue. The pony shies, Sylvia sails, and falls for a long time. They say that in dreams if you fall and then land, that means death, you are dead. In the dream, just as it happened that day, she doesn't remember landing, but only becomes aware when she is lying in the grass and cannot move. In the dream she knows she will never move again. The clouds are going by impossibly fast, the sky above is that deep-violet sky she saw in her vision of the rolling earth. If she could raise her hand, the clouds would shred through her fingers as on rough mountain peaks. In a while someone comes, the girls from the stables, to look for her. They walk past mere feet away, but do not see her. She can't hear their words, but she senses impatience, reproach in their attitude, and she doesn't care. They take the pony back, and leave her there. Darkness falls, and the clouds and the sky become a painted ceiling, and now she lies still as if on a bed, while this ceiling speeds by just inches overhead, and Sylvia knows that she will lie this way forever.

The sun on her face, sometimes shadowed by cloud. She lies down in the grass and someone walks above. The lover. He seeks her. She would call to him, but he must discover her. He circles her where she lies. He will come to her where she lies. And will he stay with her there? He cannot find her. Why? He cannot see her, and perhaps she is not there, but he will not abandon searching. He will search, she will lie, where she lies. His shadow even touches her. Is that he, or the clouds? She even hears his voice. She drifts slowly toward it, then starts up:

— . . . Bryce Fraser. I'm looking for Sylvia. I wondered if maybe she was out there, or if you'd seen her. I can't find her at her room, or at work. If you've seen her, would you give me a call. If you're the right Ted Stenmark, Sylvia's dad. You're the only one in the book. I'm Sylvia's . . . friend. Well, thanks. My number is . . .

The machine clicks off. Sylvia is left to wonder, if there had been time, would she have picked it up.

. . . *into future* . . .

Roll it into a ball, and up the hill, or up the river. Mind the fish.

Sylvia goes to tend to her laundry. Someone has taken her wet clothes from the washer and set them atop the dusty dryer. Never mind. She throws all the clothes in one machine, and pumps in quarters for an hour.

She has chores she has neglected. Water the plants. They are drooping, but they'll live. Bring in the mail. The box must be full to bursting. She finds the key and goes for the mail. Returning to the apartment she leafs through it. All bills and junk mail and insurance forms and sailing catalogs. She thinks, If I got mail like that, I'd run away to the Islands, too. She doesn't get any mail, to speak of.

She cleans up the apartment. How has she managed to make

such a mess? It seemed as if she only slept and watched TV. She puts the books back on the shelves, imagining that she is creating some kind of order among them. No one else would perceive this order. She hopes her father will notice the difference, and wonder. She hopes he will be somewhat annoyed. He won't know how to put them back the way they were. It will disturb his tidy universe, ever so slightly. She keeps out the Unamuno. She looks around the room and sees memory opposed to the spiritual life. She sees Pandora's box and Sisyphos's stone, Tantalos's garden pool. She thinks of the unspeakable and the unspoken. Of what can be proved and what must only be known. Knowing alone could perhaps exert some force. To see a thing for what it was, this might wield some subtle power.

Sylvia feels it resuming. *My life,* she thinks. What was, what is. She thinks of her room, and wonders if she can still live there.

. . . all this, I know well . . .

Passing the bookshelves, she stops, takes down Pogo. Flips it open: *We have met the enemy and it is us.* Wisdom from the mouths of opossums. Opossi? It is resuming.

She turns off the fish. The books are in their new order. Her laundry is clean, and bagged. She locks the door behind her. Leaving, she does not feel like a criminal. Nothing is clear.

Animal Rights

She had to be somewhere. Even one so evasive as Sylvia, so practiced at concealment, could not vanish into air; maybe she had gone and dissolved her elements into the earth, but I doubted it.

I had called her father's apartment, left a message on his machine. Maybe Sylvia was there, and maybe she wasn't. I wasn't up to another sortie to the suburbs.

I decided she had to be going to work. Her supervisor was in on the deception. Having given up sincerity, I was prepared to encounter duplicitousness. I almost relished the thought.

. . .

The day was overcast, likely to rain, cool and blustery. I parked in the ramp and took the elevator up to the top, went out and then came in through the doors I'd entered through on Memorial Day. I wanted to have some idea where I was in reference to the outside world, if only for a moment.

I thought it would be easy enough to find Sylvia's lab, on a regular working day, with people around. I went down a couple of levels and started assailing white-coated people in the halls, asking for Dr Langerhans's lab, but no one knew where it was. There were dozens and maybe hundreds of labs down here, each one a little suzerainty, presided over by its big boss doctor or distinguished professor, each embarked on a great and urgent destiny, raking in the grant money, slaughtering various small animals, mostly rodents, each existing in apparent ignorance of all the others.

—Langerhans, Langerhans, said a curly-headed young man with a monkey on his shoulder. —That sounds really familiar, but I don't know.

I wandered on through the halls of the great hive of science with an unsettling sense of déjà vu. I had traveled every one of these corridors just weeks before, but it felt like another lifetime. I was glad not to be alone. Eventually I would happen on Dr Langerhans's neighborhood, and someone would know where to send me. It felt a bit like traipsing through Oz. Brains and hearts could be easily had, and someone, somewhere, was surely brewing up a batch of chemical courage. Dorothy's companions had the advantage of me, for at least they knew what they needed.

I restrained the urge to skip and sing. This was really no laughing matter. But a nudging edge of the absurdity of my quest, finding myself lost again in one of Sylvia's strange terrains, made me want to laugh. The disorientation touched the strings of my secret desire and sent out a resonating note which harmonized with the mechanical hum of the complex, and made me think that maybe I really wanted to be Sylvia's knight-errant. The knight-errant went to slay the dragon and free the damsel. I searched only for Sylvia. So she was damsel and dragon both. All right. That jibed nicely with the modern psychoanalytical model.

I didn't know what I was going to say to her if I did find her. And try as I might, I couldn't imagine it. I didn't know quite why she was hiding from me, so I couldn't guess how she would react if I found her.

I turned a corner and found myself walking toward a tall gray-haired man in a fine blue suit. The wizard himself. He approached with forceful strides and his posture was superb. His manner said busy, busy, very important, and I wasn't going to trouble him, but at the last moment, just as he was about to pass, I stopped short and said, loudly,

—Excuse me! Would you happen to know where Dr Langerhans's lab is?

And he pulled up too, and turned slowly to face me, and said,

—Well, I'm Dr Langerhans.

He gave me a benign, questioning look. He wanted recognition, he wanted to know my business. My journalistic savvy emerged just in time. I extended my hand, introduced myself, said how happy I was to meet him, that I was a friend of Sylvia Stenmark and had heard so much about him. He wasn't such a straight-ahead guy as he had first appeared. In fact he seemed to relish being waylaid in the hallway—it probably never happened, with the way he walked. Like a good feudal lord he produced whatever trivia he could recall about Sylvia—when she had come to work here, where she was from—most of it imaginary, I thought. When kindly Dr Langerhans had run out of these fictional tidbits we stood looking at each other, smiling and nodding. I could tell that he was a man of encompassing intellect and drive, but he wasn't much of a conversationalist. When this grinning, bobbing pause grew awkward, I said,

—Well, I don't want to keep you. Could you point the way to the lab?

—Right around the corner, but I don't think Sylvia is in today, he said.

We said good-bye. Was he in on it too? No, not the good Dr Langerhans. I went around the corner and put my head in the first door on the right. A man stood at the center island counter with his back to me.

—Hi, Bob, I said.

He turned slowly.

—It's Robert. You're the boyfriend. Sylvia isn't here. She hasn't been here for days. I told you. Many times.

He wore the requisite white lab coat, and clear plastic gloves on his hands. One hand held a small white enameled tray, and the other a scalpel. On the tray lay six tiny, whitish, slightly bloody blobs, widely spaced. On another tray on the counter lay six mice, also neatly arranged in two rows of three, on their backs, legs in air, heads thrown back, bellies slit. I looked at the blobs on the tray Bob held, and into my mind, I don't know why, popped: *pancreases*. I strongly sensed that there was something symbolic in this scene of flayed mice, the tray of organs, butcher Bob and his scalpel, that it was some kind of omen, at least; but I couldn't fathom what it might mean. I don't think I had ever seen a pancreas before; I had never even eaten sweetbreads. I stared as if entranced. The knight faces many trials, and must learn to read the signs.

Bob just looked at me as I stood and gaped. Since he was merely an element of the omen I felt no obligation to him. He might metamorphose at any moment, take the shape of a snarling beast. But I should try to learn what I could.

—Do you have any idea where she is? I said.

—How should I know?

—Do you know why she hasn't been at work?

—Why would I?

—Can I leave a note for her here?

—I've already left messages of your myriad telephone calls.

—Sorry about that. I know I've been kind of a pain.

—You've been a royal flaming pain.

—I accept that. I apologize. You don't have to be hostile. I'm worried about her.

—I'm sure she's fine. I'm sure if she wanted to talk to you, she'd call. You have a phone, don't you? She knows your number?

His vehemence was disturbing. Was there something going on between him and Sylvia? Was she secreted away at his bachelor apartment, lounging in a negligee, eating bonbons and caviar? No. He might be in love with her, but she wouldn't reciprocate. He wasn't her type.

And what was? Was I? What type *was* I? I wanted Sylvia to see herself in pop songs; maybe I needed to do the same.

—What did she say when she called here?

—She said she wouldn't be in for a few days. As I've told you any number of times.

But I might catch him in an inconsistency. But the story was too simple. If he really didn't know anything, there was no reason for him to. And I was being a royal flaming pain.

—I'd like to leave a note, I said.

—Go ahead.

I didn't have a pen or any paper. I said,

—Do you have a pen, a notepad?

—Look, I'm busy. Now, if you'll excuse me . . .

Now, if you'll excuse me . . . he said. And he turned back to the counter and his dead mice. It was something people said on TV, and then the one so spoken to always up and left. The one who said it would pick up the office phone to take a very important call. No one said that in real life. People didn't know how to act anymore. I imagined Bob sitting alone at night in a darkened apartment, in the oracle's blue flicker, watching late-night reruns of soaps and legal dramas, and mouthing, as he squared his shoulders (or imagined that he did), *Now, if you'll excuse me . . . If you'll **please** excuse me . . .*

The person so spoken to is supposed to go away, huff off, vowing revenge. I didn't. Bob knew I was still there, he stood with his back to me. It seemed he was all finished with these mice and pancreases. He wasn't doing anything, just standing, and he was stuck. He turned to me again and said,

—*Look.* Will you *please excuse* me.

—I don't mean to bother you. I need a pen and some paper.

—Over there, by the phone, he said, pointing.

I went to the phone, found a pen and message pad. "Sylvia," I wrote in the designated space. And then the date, and time, in their proper places. These blanks were handy for guiding one's expression.

WHILE YOU WERE OUT: ________

Who was I? Your Knight Errant, Your Bewildered One, Desperate.

WHILE YOU WERE OUT: <u>**Bryce**</u>

☐ Telephoned

☐ Called to see you

☐ Wants to see you

☐ Please call

☐ Urgent

☐ Will call again

All of the above. But I didn't check anything, and continued on to:

MESSAGE:____________________

____

____

____

____

But even the neat ruled lines, the guiding title and colon, the enforced concision, could not tell me what it was I had to say to Sylvia. I stood a few seconds with my mind as empty as it could be. I began to feel self-conscious, checked over my shoulder to see Bob in the far corner. As I looked he threw a glance at me and then turned abruptly away. I bent to the message pad and rapidly wrote: *Now is the time for all good men to come to the aid of the party*—then dropped the pen on the counter, tore off the sheet, shoved it in my pocket, and left. Bob pretended not to notice. I would never be able to call here again.

I saw a cow on a treadmill, all wired up and rolling along with that uniquely bovine resignation. I heard the guinea pigs, the cats and dogs. But this time, with lab doors open, with people about, the sounds of talk and laughter, all the happy scientists going about their business, the atmosphere was not sinister but quite domestic and orderly. And I thought that although these people brought suffering and death upon their animal charges, still they must form bonds that the rest of us would never know. As with kidnappers and their victims, a powerful intimacy bloomed and flourished in this twisted hothouse. The modern self-helpers would say it wasn't love, because love is always nice and gentle in their view, you let the butterfly free and it comes back to perch on your finger. Love as benign indifference. Maybe it was just my heedless youth, but I still thought of love as a fiery tunnel vision, something aimed and focused, not fuzzy and diffuse. Oh, that's lust, they would say. That's passion, not love. But they were just playing semantic games, pretending that because there were different words for these things, then they could be winnowed away on the winds of good intentions, leaving behind only the rich chewy grain of beneficent, nurturing love. Thus we could also be rid of jealousy, spite, and duplicity. Since I was working hard at my new-found insincerity, I mentally scoffed at these good-hearted cretins. Love is a violent thing, I thought. A lot of what we do for fun is violent.

Someone passed me leading a sheep on a leash. It was like the state fair down here. I took a stroll through the halls of this level, for no particular reason. Maybe I still resisted leaving when I'd been told to, maybe I wanted to run into Bob again, have a scene, make him call security, be dragged out in a murderous rage. Maybe I still suspected that Sylvia really was here. And as I came around again to Dr Langerhans's lab, that suspicion was confirmed.

I saw her from behind, exiting the lab pushing a stainless-steel cart. I was at the other end of the corridor, and my heart leapt, and I froze. She disappeared around the corner. My heart was racing, and there was a pressure in my head that made me dizzy. Then this must be love. When I was able to move I set off at a fast walk, then began to sprint down the carpeted corridor. I called out:

—Sylvia!

And as I passed the lab door I heard Bob say,

—Jesus christ. It's—

I came around the corner and didn't see her. I dashed to the next corner, and as I rounded it I saw her turning the next corner.

—Sylvia! Wait! I yelled.

I made the next corner, into a corridor that dead-ended at the elevator, and she was at the end, pushing the button. It wasn't far, I would catch her. I ran toward her, calling,

—Sylvia! Don't—

Sylvia didn't. It wasn't Sylvia.

It was a woman, all right. About Sylvia's age and size. The similarities ended there. She had turned in alarm as I descended upon her, and stood backed into the corner by the elevator buttons.

—Hey, what're you, whadda you want?

She tried to keep an eye on me as she searched the cart top for a weapon. If there had been anything sharp there, she could have had my pancreas out in a second. I pulled up short, held up my hands.

—Sorry, sorry, I said. I thought you were somebody else.

—Well I'm *not,* she spat.

—You just, you look like her, I said.

Which she didn't. Her face was rounder, her hair much lighter, she wore large round tortoise-shell glasses. I walked slowly toward her.

—Keep away from me, she said.

—It's all right, I said. I made a mistake, I'm sorry.

And it occurred to me that lately I was apologizing to everyone, and that this was a bad sign. The woman relaxed a little. I was ready to leave now, and as long as I was at the elevator, I thought I might as well use it. I tried to assume a nonchalant, waiting-for-the-elevator posture. If I hadn't been so short of breath I might have whistled a breezy tune. The woman watched me out of the corner of her eye. The elevator dinged. We both jumped. I said,

—Going up?

—Down, she said.

—I'll wait for the next car.

The doors opened and she rolled her cart on backwards, watching me as she did so.

—Sorry, sorry to have scared you, I said. Bye.

I waved. She did not. The doors closed on her. Someone else I'd made angry with me. Irene had said I made a good first impression.

And as I rode upward in the elevator, I told myself: When you find yourself doing harm, stop what you are doing. And I resolved to stop.

Outside, the sky was spitting rain, and there was a demonstration going on. A dozen or so people with signs—and one with a tambourine—circulated in what at first appeared some kind of odd dance. The flat area before the door was not large enough for their circling picket, and they were forced to climb and descend sets of three steps at opposite sides of the lab entrance. I had only a moment to take this in before a great round boulder of a man came swaying rapidly toward me, thrusting a handful of pages at me with foreshortened arm. The pages were photographs.

—How can you do this to your fellow creatures? he said, waving the photos.

He presented the top photo for my perusal. It showed a monkey's scorched face, teeth clenched in an agonized death grin.

—I'm just visiting, I said.

—Down there, in that veritable hell, they're torturing puppies and kittens.

—Mostly it's rodents, I said. Some livestock.

The protesters were chanting something, but they all seemed to be chanting something different.

—They're peeling off their skin, they're burning them with acid. Don't you care?

I almost said, not really, better them than us. I had to wonder how the man had grown so enormous without the aid of animal products. I said:

—I don't work here. I don't have anything to do with it.

I who had stood idly by while countless mice were depancreated, who had looked on, impassively, and done nothing.

—Next time you use shampoo, the man spluttered,—think of all the rabbits that suffered and died.

I thought it very, very unlikely that I would do that.

—Next time you put on make-up, think about that.

The man was a fanatic, and also a lunatic. When he wasn't accosting nearly innocent non-scientists he wrote letters-to-the-editor that were never printed and ate copious quantities of junk food. He would go on a shooting spree one day, and when he came to trial his defense would be that he was driven insane (temporarily) by man's inhumanity to lab animals. Briefly I felt afraid. What if they were all like him, desperate animal-rights terrorists, and having waited so long to face their nemesis (were they stalking my dear friend Dr Langerhans?) they chose to seize and flay me in retaliation, and my life would end with the grisly, ironic denouement of a 1930s novel, just to show that men and women are beasts, after all, so what do you expect? But when I surveyed the crowd I saw them watching us with obvious embarrassment. They were ordinary, decent-looking people. While they watched us they weren't concentrating on the dicey business of split-level marching, and they tripped and stumbled on the steps, and their circle moved jerkily along, piling up at the stairs, stringing out on the flats. A University security officer stood a little distance off. He fingered a nightstick. He checked his watch, and then the sky. And he was watching me and my interlocutor, too, and I couldn't tell if he did or didn't want something ugly to happen. Maybe on another day, when the weather was better. I started to move away from the huge man, and he shook his finger and said,

—Abortion kills babies!

But I made my way around him—I had to move literally around him, a crowd unto himself—and around the rising and falling oblong of marchers. None of them bothered me, though none apologized for their comrade's unseemly manner, either. The rain fell harder as I crossed the avenue on the pedestrian bridge to the main campus.

As I walked across campus in the rain I kept telling myself, *This is it, this is the last time.* Like an alcoholic who tells himself that every drink is his last, maybe. How is it that we can tell ourselves something, and know we're lying, and yet still believe it? A function of the secret desire: if you say to yourself *a, a, a,* but *b* is what you really want, *b* it will be. But I was nearing the point where desire shifts. Having wanted and striven after *a* for so long, now perhaps it was *b* that I really wanted. It's all completely selfish. *a* is to have Sylvia, and *b* is not to have her. And *a* is no more noble than *b*. When one is truly forced to confront one's motives, it is difficult to believe in nobility at all, or in selflessness, or true charity, or maybe even in love.

But these thoughts arose as I was being drenched in rain, having assaulted one person and having been assaulted by another (a taste of my own medicine), on my way to look (one last last time, I promise) for Sylvia whom I would not find; and so they were contingent.

Sylvia did not exist. I could believe this as easily as not. I had imagined her into being, but an imaginary being, though she may be real in some sense, is not flesh, cannot be held and loved. I climbed the fire escape. I would not find her here, then why did I look? It made no sense, it had no point; but can't we get beyond that, at last? The sense, the point, the reason: these are fictions also. I was going to look one last time for Sylvia because that was what I had decided to do. If I were wrong, and it turned out she was there in her room, and greeted me and welcomed me in as if nothing at all had happened, that would not provide a point to my seeking, it would not change anything.

Sylvia's room was dark, empty. The rain dripped down off the eaves. I was already drenched, but sheltered here outside Sylvia's door. Sylvia had been here: the bed was made—Mr Somebody tucked and hidden. And one other change since my last visit: the list of house rules had been torn from the door; the thumbtacks held their corner scraps of paper like worried mothers clutching handkerchiefs while the police drag the river. She had trashed "The Last Supper"; she had renounced Mrs Simic's commandments. It meant something, but how did I know what it meant to Sylvia? I looked in at the tiny dark room, and I thought of the cedar closet. Was that the locus of meaning in her life? Did her secret desire drive her to re-create the place itself, and the feelings of betrayal, isolation, desertion, and outrage that it must signify for her? I wondered then if the secret desire wasn't precisely focused on reliving some such crucial moment, perhaps in order to set right what was long wrong. But did you ever set it right? Sylvia would never admit that this was so. She would scoff at the very idea of the secret desire.

So Sylvia had been here, there was evidence of Sylvia. At that moment I ceased to wonder or even care where she was. I was done with hide-and-seek. I was not so foolishly in love, or obsessed, that I would go on searching for Sylvia while she fled from me. Then I thought of her night in the cornfield, and maybe she still wanted Father to find her at last and carry her to warmth and safety. These paradigms of behavior were becoming too neat, the secret desire was coming to seem too clinical, and somehow tainted. I didn't want to be her father, or her confessor, or her psychologist. I just wanted to be with her. I still wanted that.

I turned from the windows and squatted on the landing. The rain had eased somewhat, the sky was still roiling black. I lit a cigarette with reluctant damp matches. A sound drew my attention up to the eaves. Two pigeons clung to a ledge to my left. They fluttered and gurgled, a liquid gulping coo. They cocked their smooth round heads and blinked their bright black eyes. It was almost cozy, the three of us under the eaves. I felt like a pilgrim, arrived to find the shrine defunct. A skeptical pilgrim, with few expectations. Now the only thing to do was to go home, find another place to ply my devotions, or just give it up. Find a hobby.

As I stood to leave the pigeons also departed, arcing away in downward flight, trailing cries of soft alarm.

At the second floor I paused and looked in through the tall arched window. I had long ceased ducking down to avoid being seen, but I had never stopped to look in. No reason to, really.

This day I stopped and looked, maybe because it didn't matter anymore, maybe out of some impulse to avert the tragic cinematic coincidence of having Sylvia walk unseen past the window and down the hallway while I descended the dripping fire escape, sunk in my despairing thoughts, unobservant (I even imagined the shot: from outside, me in the rainy foreground, dripping hair, water streaming down my face, and down the windowpanes, with her behind, in the warm, yellow-lighted hallway, turning onto the landing, walking away down the hall as I disappear at the bottom of the frame—but why is she wearing a '40s floral dress in aqua and rose, seamed stockings, black heels, and clutching a pearled evening bag?).

And when I looked, there was someone there. There was no question of mistaken identity this time. The woman in the hallway was not Sylvia, but Mrs Simic. This was somehow more astonishing than had it been Sylvia. She stood at the far end of the hall, at the foot of the stairs, hand on the newel post, looking up. It was funny, her perpetual black gabardine very likely hailed from the '40s. Her stockings indeed were seamed—be careful what you wish for, it was my *film noir* darling gone to seed. This was such a novel sight, Mrs Simic on the second floor, contemplating the third, I simply stopped and stared. Her head was directed upward, as if watching or listening for something. She wasn't saying anything, for her mouth was pinched in a worried frown. She was about the same age, and of the same build, as Sylvia's grandmother, and as I watched her there came another vision from the early days of Sylvia, Gran flaming to her god, and I thought, well then this is how stocky old women go to God, not flaming like angels but clumping up to heaven in orthopedic shoes, with a rest at the mezzanine. If I were a painter I would have rendered the final ascent just thus, and I would include the dog, Kaiser.

And then she turned her face to the window. Our eyes met.

Did they? I can't say for sure that she saw me at all. It was quite dark outside, late afternoon and heavy skies, and it was bright within. Her eyes were poor, her glasses so thick it was hard to tell what she was seeing. And the window was streaked with rain, and only my head and shoulders would have been visible. If she saw anything it was only a dark shape at the window, and maybe she wasn't sure if there was anyone at all, whether she was watched or alone, and was trying not to embarrass herself with undue alarm. She was aware of something she hadn't been aware of before, and if she saw me she made no sign, only stood with her plump gray hand on the ball of the newel post with her face turned to the window. Most likely her focus failed somewhere in the air between us, yet I had the sense of being utterly and nakedly perceived. Who saw me?

She raised her head again toward the third floor. I continued quickly down the wet fire escape, and I thought that she must remain there, a statue of doomed vigilance, forever.

Our Lady of the Corn

Rare now is the river that wanders at its own sweet will, Wordsworth. Every navigable way is dammed, diverted, channeled, stopped with locks, drained or dispersed.

That innavigable sea, the antagonist of bliss, is the only wild ocean.

I finally got around to *Two Gentlemen of Verona.* I read it sitting in Frank's studio, and I told him the convoluted tale as it unfolded. Silvia in fact is not a minor character, just one of the less interesting principals, the daughter of the Duke of Milan, a drop-dead beauty—auburn locks, gray eyes, low forehead.

—Silvia is loved by three: there's Valentine, our hero, and

his chief rival, Thurio, and then Valentine's erstwhile best friend, Proteus, who's supposed to be betrothed to Julia.

—Betrothed, what an excellent word. Thurio? He'll never make it, with a name like that. Thuringer.

—No, and he's the one her father wants her to marry. He's a goner.

Valentine, who wasn't looking for love, is pathetic when he falls. His valet Speed, who gets all the good lines, tells him that his follies shine through him "like the water in an urinal." Valentine is stricken by Silvia's beauty, and Speed asks if he's seen her since she was "deformed." *Val.:* Since when is she deformed? *Speed:* Ever since you loved her. *Val.* (puzzlement, consternation): I've loved her since I saw her; she looks great. *Speed:* Since you love her, you can't see her.

Julia, disguised as a boy, comes looking for her beloved Proteus. She makes a comely lad. I caught something a little electric in her encounters with Silvia.

—Valentine, through Proteus's scheming, runs afoul of the Duke, and he's banished. Out in the woods he's set upon by thieves, but they're honorable thieves, and he becomes their captain, like Robin Hood. They don't molest women or poor people.

—Why can't we have thieves like that?

—Silvia refuses Thurio—

—Naturally.

—and Proteus. And goes off to find Valentine.

The usual stew of mistaken identity and misperceived intentions, the lovers all suffering and equivocating, pining, plotting. Parental problems, friends betraying friends. Why didn't they just *stop?* But of course they persevere, in their bumbling way, and all the strife and turmoil end "with triumphs, mirth, and rare solemnity," and the promise of a grand wedding feast.

The scene of Launce with his dog in the fourth act really is a treat.

I knocked on my neighbor's door to give him back the book. He answered looking disheveled as usual. And his apartment looked even more overrun with books than before. He still said I should keep the play, but I insisted otherwise. He was leaning in the door as I began to turn away.

—All right. How'd everything go with Sylvia?

I turned back.

—What?

—With Silvia, did everything turn out okay?

—I thought you said Sylvia.

—Silvia, right. Noble fair Silvia. Happily ever after?

—Of course.

The drive to Gran's took no time at all. I wasn't aware of a single turn, stop, lane change, or exit. The car took itself home. Enough of the open road. It was an old car, after all.

There was a brand-new car in the drive. I pulled in behind it. The new car was an American subcompact, blue again. It gave off a soft blue shine under the light through marbled clouds. I felt betrayed, on behalf of the Bel Air, and somewhat scornful—the new car was graceless, ugly, a squashed little box of a car, with tiny little tires. How the Bel Air must have shone in 1962! But would never shine again, the finish weathered dull, gray undercoat rising to the surface. Maybe now one of the grandkids would get the blue Bel Air. And god help him, or her. But maybe its dire magic only worked on unstable minds, like mine.

As I cleaned out the interior, gathering up empty cigarette

packs and pop cans, candy wrappers and snack bags—the detritus of my all-night drives—emotion began to swell in me: reluctance to leave the car, to say this is the end, to renounce this last tie to Sylvia. Gran didn't need this car, now that she had a new one. It would be a burden to her, another piece of trash, if I left it here . . .

No. It was taking the expedient course that had brought me to this, that had led me through all this.

Was that so bad? Maybe taking the car had been a truly intuitive act, a branch of the secret desire.

Be that as it may.

I got out of the car and walked toward the garage to deposit the trash. Kaiser came skulking, wriggling out to meet me, her chain ringing with an uncanny blend of eagerness and fear. Callously I shooed her back. I dumped the trash and started back down the drive. Passing the dog's hiding place, I weakened. I bent to one knee and called to the dog, softly,

—Kaiser, come here sweetheart. Come on.

She inched forward on her belly, then stopped well beyond my reach. She looked around with frightened eyes, and would approach no nearer.

The interior of the car was now tidy if not exactly clean. I checked under the seats, then opened the glove compartment. Inside there was a rag, and a tire gauge, and an ancient map. These belonged to Gran. At the back of the glove compartment I found a roll of film. At first I thought it was a roll I had shot for the newspaper, thrown in and forgotten. But it was color film, not the black-and-white stock I got from the paper. I was sure it wasn't Gran's. I pocketed it.

Then there was no more delaying. I slammed the front door for the last time—what a sweet, solid sound. I went to the trunk, which I had never used, but I thumped it twice, as Gran had shown me, just to see if I could do it. It popped up open-sesame, and there inside was Grandpa Bastard's gray overcoat. Bought and paid for. I closed the trunk and left it there. It really didn't fit me, or so I hoped.

The house and yard were quiet, no sign of Gran. The front yard was a mess, as I had left it in my attempt to bring order there. It looked like a battlefield. A battle of wills. A holy war. The will of weeds, triumphant. The weeds always win. They had grown back tall enough to conceal the stack of wood from the apple trees.

In an upstairs window the wind softly urged white curtains inward. Maybe Gran was napping there. Or maybe if I ventured around back I would find her splayed stiff and lifeless at the verge of the creeping weeds, her hands around the reaper's scythe, its blade reddened only by rust and weed spittle; dropped dead in the quick, like the ravening shrew, as she desired.

I walked to the end of the drive. At the road I turned around. Everything about Sylvia, about me and Sylvia, from beginning to unfinished end, was represented in some aspect of this scene. The old car, the blue Bel Air, completed the composition, and so must be relinquished to it. I toyed with the phrase: blue—sad, melancholy; bel, bella, bellissimo—the continental touch, something lovely; air—a song, an atmosphere. Beauty dwells in sadness, didn't I say?

Why couldn't it have been a Biscayne, an Impala?

I was all unstrung, and phrases I couldn't control careered through my mind: *I never even had a key, I never had the key.*

But I didn't need a key.

And bel this and blue that. My blue belle. My blue belle's air. The lovely sad song of my blue belle.

Why couldn't it have been white, gold, black?

It begins with a poem about things that exist and do not. But even the calculus of imaginaries must end in the world of real numbers.

I started down the road, and I did not look back. I walked without what you could really call thinking. These phrases about keys and weeds and belle's blue airs bounced around my mind and marched in time with my stride, and I didn't know if they were nonsense syllables or distillations of purest import.

I passed new houses under construction. I picked up a stone and threw it into an open foundation and heard it ricochet and echo around the concrete walls and floor. The houses were going up but they might as well be falling down. They would go up, they would fall down. Let us sing of impermanence: all things swim and glitter, all fall down. In the end would be weeds—and Gran's house standing once again amid wilderness. But I didn't look back. How could I be sure it stood, even now?

I hadn't given any thought to how I would get home. If I didn't show at work in a few days, they might send out a search party, and find me wandering senseless in the subdivisions, or face-down in two inches of dirty water, like some Alzheimer's victim who takes off on one final fatal hegira. You could disappear without hardly trying. But I didn't want to disappear. I wanted, in fact, to get home. In my pocket I had the roll of film I had found in the glove compartment, and I was curious about it. For no good reason I felt that it held a shred of hope in its canister filled with silver and captured light.

I guessed I could catch a bus, if buses came out this far. I could hitchhike. There was traffic, although sparse. I stuck my thumb out for the first car that passed, and miraculously, it stopped. A little station wagon with two teenage girls in it. Picking up strange men on the roadside, living dangerously. I opened the back door.

—Hi, said the driver. Where're ya goan?

—I'm trying to get back to the city, I said. Do you know where I can catch a bus?

The two conferred. They had never ridden a bus before. But they had seen buses, up at the mall. Maybe those ones went to the city. There was like a park 'n' ride thing there. Anyway, they were going to the mall.

—Hop in, they said in unison.

I hopped in. They were listening to a pop radio station at a volume that precluded conversation, though occasionally one would shout something to the other. The car had an excellent stereo that made me long for the Bel Air's AM radio.

They dropped me at the park 'n' ride place in a distant corner of the mall's oceanic parking lot. You would almost want to take a bus to the mall from here. There were four cars parked here like exiles standing longingly at the frontier of their native land. I checked the schedule and saw that I had just missed a bus, and there wasn't another one for over an hour. The roll of film still nudged my leg. I thought I might find a one-hour photo place in the mall, have the film developed, and still make the bus.

I dropped off the film and then tried to amuse myself in the mall, but I couldn't even pretend to be interested in anything I saw, and time, as they say, seemed to stand still. I kept winding up back at the photo place, loitering outside the door, coming in to see if my pictures were ready, making the clerk nervous. I tried to act cheery with her, but she could see I was desperate and possibly deranged. On my fourth return the pictures were ready, and the clerk was perhaps more relieved than I when she handed them over and took my money. I scurried away without my change, and she shouted after me, and I retrieved it and left with apologies.

I sat at an unsteady round table in the mall's fast-food heaven. I had bought a cup of horrible coffee, the kind of coffee that makes you wonder why people drink coffee at all. I lit a cigarette. I was trembling and I handled the packet of photos with sweaty hands, as if it were some dreaded piece of mail I couldn't bring myself to open. Mind you, I had no idea what these pictures were. But they had been in the blue Bel Air, and so they were charged with omen.

Then I fumbled with the gummed flap and the table teetered and my coffee slopped. I jerked out the photos and the negatives flew everywhere, into the puddles of coffee on the table, onto the sticky floor, and as I grabbed for them the pictures fell into my lap. Ineptly I gathered them up, dropped a few, patiently persevered. I went for napkins. I was the only customer in the "food court," and I was acutely conscious that the teenage drones at all the food stands had nothing to do but watch my distress. I returned to my table and slowly, carefully mopped up the coffee, dabbed at the negatives which were now sticking to each other. I took the soggy napkins and the coffee cup and threw them out. I sat down and breathed.

The first picture was a blurred dark shot of pavement and someone's feet. Sylvia's feet, in white sneakers. A mistake, the beginning of the roll. The next was of a church and a graveyard on a hill. I remembered, I relaxed. The pictures from the Best Angel Contest followed: Sylvia, Carla, and me posing behind the headless angel in the derelict graveyard. The angel was not quite life-size, and so we contorted our bodies to hide behind it. And our heads looked rather grotesquely large, more like animate gargoyles than angels. Sylvia had won the contest, but as I looked at the pictures I concluded that Carla and I had given it to her. Carla made much the better angel, with her

Semitic features, her golden-red hair luffed out in a glowing aureole. Sylvia's expression was more pained than mystic. I simply looked like an idiot.

Some pictures from the farm were mixed in with random shots around the cemetery. Carla had even gotten one of the well-driller at his most irate, standing at the bottom of the hill with his mouth wide open, his arms upraised. I laughed in the empty food court, and didn't care if every teenager in suburbia thought I was insane. I saw the farmhouse, the river, the bridge of betrayal, as I had imagined it that day.

Only one really surprised me, a shot from the cemetery. Sylvia and I lie in the long grass. It is after the angel contest when we were rolling in the grass laughing. But here we are still. We lie side by side, on our backs, Sylvia on the right. Our heads are turned toward each other, and on our faces are calm smiles, as if we've just recovered from a laughing fit. We're looking at each other through the stalks of grass, each of us making a separate relief. The photo shows us from the waist up. At the very bottom you see my left hand, open and relaxed, and Sylvia's right hand resting in it.

You would swear we are in love.

The smell of crushed leaf and seed, a memory scent, overwhelmed the cloying mall smell of fastfood and new clothes. The picture, and our embrace in the weeds that first time at Gran's, and that night that I held her and told her I loved her and wouldn't let her go—they all fell together, as if it had all been one instant. I could feel her hands on my back, as if all sensation I had ever known was channeled through her small hands. I could smell the scent of the dry field, and the scent and the heat of Sylvia in bed, and I felt the heat of the sun. How could it be over?

The mall was freezing. It smelled horrible in here. I put

away the pictures, and I threw away the ruined negatives, and I left.

I'm alone on the bus, which eschews the freeway for a local route. The rows of corn and the path of the bus conspire to contrive a vertiginous geometry, like domino rows of gravestones. In this flickering vision I see, in the center of the field, crouched low and small, Sylvia, tiny Saint Sylvia, Our Lady of the Corn. She is dissolving her elements into the earth, and each time I glimpse her she grows less distinct. The dissolution proceeds with gathering pace. She no longer needs sense or right reason. No one is calling her name, or if someone is, she doesn't hear him. She is becoming diffuse and integral, something better than truancy, than invisibility, even. She is becoming infinitesimal bits, without ego, without memory, without bounds. She digs in her toes—she just barely has toes—and for an instant does she pause to reconsider? No, this is it: she is going, she is gone.